IMPROVING THE MEASUREMENT OF LATE-LIFE DISABILITY IN POPULATION SURVEYS

Beyond ADLs and IADLs

Summary of a Workshop

Gooloo S. Wunderlich, *Rapporteur*

Committee on National Statistics
and
Committee on Population
Division of Behavioral and Social Sciences and Education

NATIONAL RESEARCH COUNCIL
OF THE NATIONAL ACADEMIES

THE NATIONAL ACADEMIES PRESS
Washington, D.C.
www.nap.edu

NATIONAL ACADEMIES PRESS 500 Fifth Street, N.W. Washington, DC 20001

NOTICE: The project that is the subject of this report was approved by the Governing Board of the National Research Council, whose members are drawn from the councils of the National Academy of Sciences, the National Academy of Engineering, and the Institute of Medicine. The members of the committee responsible for the report were chosen for their special competences and with regard for appropriate balance.

This study was supported by Contract No. N01-OD-4-2139, TO # 195 between the National Academy of Sciences and the National Institute on Aging. Support of the work of the Committee on National Statistics is provided by a consortium of federal agencies through a grant from the National Science Foundation (Number SES-0453930). Any opinions, findings, conclusions, or recommendations expressed in this publication are those of the author(s) and do not necessarily reflect the views of the organizations or agencies that provided support for the project.

International Standard Book Number-13: 978-0-309-14371-4
International Standard Book Number-10: 0-309-14371-3

Additional copies of this report are available from the National Academies Press, 500 Fifth Street, N.W., Lockbox 285, Washington, DC 20055; (800) 624-6242 or (202) 334-3313 (in the Washington metropolitan area); Internet, http://www.nap.edu.

Printed in the United States of America

Suggested citation: National Research Council. (2009). *Improving the Measurement of Late-Life Disability in Population Surveys: Beyond ADLs and IADLs, Summary of a Workshop*. Gooloo S. Wunderlich, Rapporteur. Committee on National Statistics and Committee on Population. Division of Behavioral and Social Sciences and Education. Washington, DC: The National Academies Press.

ERRATA

On page viii, the second sentence in the last paragraph should read:

The review of this report was overseen by Eileen M. Crimmins, Davis School of Gerontology, University of Southern California.

Improving the Measurement of Late-Life Disability in Population Surveys: Beyond ADLs and IADLs
The National Academies Press, Washington, DC, 2009
ISBN-13: 978-0-309-14371-4
ISBN-10: 0-309-14371-3

THE NATIONAL ACADEMIES

Advisers to the Nation on Science, Engineering, and Medicine

The **National Academy of Sciences** is a private, nonprofit, self-perpetuating society of distinguished scholars engaged in scientific and engineering research, dedicated to the furtherance of science and technology and to their use for the general welfare. Upon the authority of the charter granted to it by the Congress in 1863, the Academy has a mandate that requires it to advise the federal government on scientific and technical matters. Dr. Ralph J. Cicerone is president of the National Academy of Sciences.

The **National Academy of Engineering** was established in 1964, under the charter of the National Academy of Sciences, as a parallel organization of outstanding engineers. It is autonomous in its administration and in the selection of its members, sharing with the National Academy of Sciences the responsibility for advising the federal government. The National Academy of Engineering also sponsors engineering programs aimed at meeting national needs, encourages education and research, and recognizes the superior achievements of engineers. Dr. Charles. M. Vest is president of the National Academy of Engineering.

The **Institute of Medicine** was established in 1970 by the National Academy of Sciences to secure the services of eminent members of appropriate professions in the examination of policy matters pertaining to the health of the public. The Institute acts under the responsibility given to the National Academy of Sciences by its congressional charter to be an adviser to the federal government and, upon its own initiative, to identify issues of medical care, research, and education. Dr. Harvey V. Fineberg is president of the Institute of Medicine.

The **National Research Council** was organized by the National Academy of Sciences in 1916 to associate the broad community of science and technology with the Academy's purposes of furthering knowledge and advising the federal government. Functioning in accordance with general policies determined by the Academy, the Council has become the principal operating agency of both the National Academy of Sciences and the National Academy of Engineering in providing services to the government, the public, and the scientific and engineering communities. The Council is administered jointly by both Academies and the Institute of Medicine. Dr. Ralph J. Cicerone and Dr. Charles M. Vest are chair and vice chair, respectively, of the National Research Council.

www.national-academies.org

STEERING COMMITTEE ON IMPROVING SURVEY MEASUREMENT OF LATE-LIFE DISABILITY: BEYOND ADLs AND IADLs

ALAN M. JETTE (*Chair*), Health & Disability Research Institute, Boston University
VICKI A. FREEDMAN, School of Public Health, University of Medicine and Dentistry of New Jersey
LINDA P. FRIED, Mailman School of Public Health and College of Physicians and Surgeons, Columbia University Medical Center
LINDA G. MARTIN, Senior Fellow, RAND Corporation
JOSHUA SALOMON, School of Public Health, Harvard University
ARTHUR A. STONE, Department of Psychiatry and Psychology, Stony Brook University

GOOLOO S. WUNDERLICH, *Study Director*
BRIDGET EDMONDS, *Administrative Assistant*

COMMITTEE ON NATIONAL STATISTICS
2009

WILLIAM F. EDDY (*Chair*), Department of Statistics, Carnegie Mellon University
KATHARINE ABRAHAM, Department of Economics, University of Maryland, and Joint Program in Survey Methodology
ALICIA CARRIQUIRY, Department of Statistics, Iowa State University
WILLIAM DuMOUCHEL, Phase Forward, Inc., Waltham, MA
JOHN HALTIWANGER, Department of Economics, University of Maryland
V. JOSEPH HOTZ, Department of Economics, Duke University
KAREN KAFADAR, Department of Statistics, Indiana University
DOUGLAS MASSEY, Department of Sociology, Princeton University
SALLY C. MORTON, Statistics and Epidemiology, Research Triangle Institute, NC
JOSEPH NEWHOUSE, Division of Health Policy Research and Education, Harvard University
SAMUEL H. PRESTON, Department of Sociology, University of Pennsylvania
HAL STERN, Department of Statistics, University of California, Irvine
ROGER TOURANGEAU, Joint Program in Survey Methodology, University of Maryland
ALAN ZASLAVSKY, Department of Health Care Policy, Harvard Medical School

CONSTANCE F. CITRO, *Director*

COMMITTEE ON POPULATION

LINDA J. WAITE (*Chair*), Department of Sociology, University of Chicago
EILEEN M. CRIMMINS, Department of Sociology, University of Southern California
PETER J. DONALDSON, Population Council, New York
BARBARA ENTWISLE, Department of Sociology, University of North Carolina
JOSHUA R. GOLDSTEIN, Max Planck Institute for Demographic Research, Rostock, Germany
CHARLES HIRSCHMAN, Department of Sociology, University of Washington
BARTHÉLÉMY KUATE-DEFO, Department of Demography, University of Montreal
WOLFGANG LUTZ, World Population Program, International Institute for Applied Systems Analysis, Laxenburg, Austria
DUNCAN THOMAS, Economics Department, Duke Global Health Institute, Duke University
BARBARA B. TORREY, Population Reference Bureau, Inc., Washington, DC
MAXINE WEINSTEIN, Center for Population and Health, Georgetown University

BARNEY COHEN, *Director*
JACQUELINE R. SOVDE, *Senior Program Assistant*

Acknowledgments

This report is a summary of the January 2009 workshop, *Improving the Measurement of Late-Life Disability in Population Surveys: Beyond ADLs and IADLs*. The workshop was convened by the Committee on National Statistics (CNSTAT) in collaboration with the Committee on Population, both in the Division of Behavioral and Social Sciences and Education (DBASSE) of the National Research Council (NRC). Support for the workshop was provided by the National Institute on Aging.

As chair of the steering committee to plan the workshop, I wish to thank my colleagues on the committee for their helpful guidance and leadership in planning the workshop and moderating the sessions.

I acknowledge with appreciation the many people who participated in the workshop and contributed to its success. I thank all the presenters for their expert presentations and Barbara Altman for a comprehensive background paper. I also thank all the participants for their stimulating and insightful comments and discussion.

On behalf of the steering committee, I acknowledge with gratitude the contributions of staff. Constance F. Citro, director of CNSTAT, provided guidance and support throughout the project. Gooloo S. Wunderlich, study director for the workshop, did an outstanding job of overseeing the detailed planning of the workshop based on the guidance of the steering committee, as well as preparing this summary and responding to the comments of the reviewers. Bridget Edmonds managed the administrative details and workshop arrangements. Eugenia Grohman, associate director for reports, DBASSE, and Christine McShane, senior editor, DBASSE, provided professional editing advice, and Kirsten Sampson-Snyder efficiently shepherded the report through the report review process.

This report has been reviewed in draft form by individuals chosen for their diverse perspectives and technical expertise, in accordance with procedures approved by the Report Review Committee of the NRC. The purpose of this independent review is to provide candid and critical comments that will assist the institution in making its published report as sound as possible and to ensure that the report meets institutional standards for objectivity, evidence, and responsiveness to the study charge. The review comments and draft manuscript remain confidential to protect the integrity of the deliberative process.

We thank the following individuals for their review of this report: Arun Karlamangla, Geriatric Medicine, University of California, Los Angeles; David W. Keer, Office of Special Education and Rehabilitative Services, U.S. Department of Education; Robert A. Pollak, Olin Business School, Washington University; and Robert B. Wallace, Department of Epidemiology, College of Public Health, University of Iowa.

Although the reviewers listed above provided many constructive comments and suggestions, they were not asked to endorse the content of the report nor did they see the final draft of the report before its release. The review of this report was overseen by Eileen M. Cramming, Davis School of Gerontology, University of Southern California. Appointed by the NRC, she was responsible for making certain that an independent examination of this report was carried out in accordance with institutional procedures and that all review comments were carefully considered. Responsibility for the final content of this report rests entirely with the author and the institution.

Alan M. Jette, *Chair*
Steering Committee on Improving
Survey Measurement of Late-Life
Disability: Beyond ADLs and IADLs

Contents

1

Introduction

This report is a summary of a workshop organized to draw upon recent advances to improve the measurement of physical and cognitive disability in population surveys of the elderly population. Are the measures of activities of daily living (ADLs) and instrumental activities of daily living (IADLs) used in many population surveys sufficient as the primary survey-based indicators of late-life disability? If not, should they be refined or should they be supplemented by other measures of disability in surveys? If yes, in what ways should disability measures be changed or modified to produce population estimates of late-life disability and to monitor trends. What further research is needed to advance this effort?

BACKGROUND

People in the United States today are living longer and healthier lives than ever before. The average life expectancy at birth increased from 54.8 years in 1915 to 68.2 years in 1950 to an estimated 77.8 years in 2005. The average life expectancy at age 65 has increased from 77.8 years in 1950 to 83.7 years in 2005 (He et al., 2005; National Center for Health Statistics, 2009). While chronic conditions increase with age (about 80 percent of people aged 65 years and older have at least one chronic condition and 50 percent have at least two), and so do disability and functional limitations, there has been a downward trend in the estimated prevalence of disability among the older population.

The most commonly used measures for estimating trends in disability from population surveys have been ADLs and IADLs. These measures were developed several decades ago in response to the recognized need to assess functional status and disability rather than relying solely on measures of self-perceived health or data on previous health histories. ADLs measure people's abilities to perform basic tasks of daily life without assistance, including eating, dressing, bathing, using the toilet, and getting in and out of bed. IADLs measure people's abilities to function independently in carrying out such activities as housework, preparing meals, shopping, managing money, and using the telephone. Questions about ADLs and IADLs are easy to administer to obtain self-reports or reports from caregivers. Estimates of the number and characteristics of older adults with difficulties in performing ADLs have important implications for health policies and programs and associated costs of health care for the elderly population.

Data on ADLs from the National Long-Term Care Survey (NLTCS) showed a marked decline in prevalence rates of disability among the elderly population from 25 to 19 percent over the period 1984 to 1999, an annual decline of nearly 2 percent per year from 1984 to 1999 (Manton et al., 1997; Manton and Gu, 2001). Researchers have identified some of the factors contributing to the decline: improvements in medical treatment including new and improved medicines and technology, improvements in health behavior, increased use of assistive devices to help people cope with impairments, environmental changes, changes in socioeconomic status and exposure to disease, and increase in educational attainment and related changes in occupational composition among older people.

Other research has broadly supported those findings, although with variations in the estimates of prevalence of ADLs or IADLs. However, not all studies have found declines for all measures (see, e.g., Crimmins et al., 1997; Waidman and Liu, 2000; Schoeni et al., 2001; Freedman et al., 2002; Freedman, 2006). Although the focus of this workshop is disability measurement in late life, note that trends in reported disability rates among younger cohorts, especially those 30–50 years of age, appear to be rising even as the elderly are becoming less disabled (see, e.g., Lakdawalla et al., 2004; Soldo et al., 2006).

Because of these inconsistencies, several efforts have been undertaken to examine the estimated prevalence rates and trends in disability. In 1988 the Federal Interagency Forum on Aging Related Statistics appointed a committee to systematically examine the different estimates from the various surveys measuring ADLs. The committee found substantial differences in the prevalence of disabilities for the elderly population across the various national surveys. Sources of variation included diversity in conceptual definitions of disability, the nature of reference periods used for recall, the disability measures used, the wording of questions, the sampling frame, the reporting rules for answering ADL questions, the order of the questions on the questionnaire, the modes of data collection (in person, by telephone, etc.), the age composition of the respondents, and the threshold levels used for construction of disability statistics and other features of the analysis (U.S. Department of Health and Human Services, 1989; Wiener et al., 1990).

About a decade later, the report of a workshop (National Institute on Aging and National Bureau of Economic Research, 2001) suggested that work was needed to reconcile different disability measures, such as chronic illness rates, functional limitations, ADLs, IADLs, uses of support services, and physical versus cognitive health; to craft new or revised survey questions to increase comparability across surveys; to test actual physical performance (e.g., climbing stairs) in addition to answering questions; and to develop measures that assess people's ability to live independently, considering objectively measured physical and cognitive abilities and assistance from devices and services that foster independent living even without improvements in underlying health.

In September 2002, a technical working group, funded by the National Institute on Aging (NIA), considered how to resolve inconsistencies in old-age trends in ADLs across national surveys (Freedman et al., 2004). That was followed by the report of a workshop convened by the Office of the Assistant Secretary for Planning and Evaluation in the U.S. Department of Health and Human Services on improving survey measures of late-life disability, which also discussed needed work in this area (Freedman et al., 2005).

In addition to the issue of the inconsistencies of findings based on ADLs and IADLs is the question of whether ADLs and IADLs continue to be sufficient as survey-based measures of disability. Should they be refined or augmented with other measures that could better track progressive declines in functional status? A related question is how responses to ADLs and IADLs may have changed over time because of changes in the environment, such as availability of assistive devices, health care, and other social services that affect respondents' abilities or perceptions of their abilities.

Other possible methods are now being developed for policy-relevant disability measures in addition to ADLs and IADLs that are feasible to administer in population surveys. One way is afforded by time-use survey methodology to refine or augment existing measures of ADLs and IADLs, ranging from 24-hour diary surveys to more intensive measurements of activity. Time-use surveys can capture the frequency and time spent on activities, such as getting dressed, exercising, driving, and shopping. Such methods offer ways to validate and improve other measures and may provide additional measures for standardization across surveys. Performance measures of functional limitations, such as chair stands, gait speed, and grip strength, could be included in surveys, and their relationship to current self-reports of ADLs and IADLs could be analyzed.

In addition to new methods to measure disability, other changes in survey methods offer opportunities for measuring and understanding disabilities. One such change is in the mode of surveys: from personal interviews using paper-and-pencil questionnaires to telephone interviews to computer-assisted personal interviews and, most recently, to the Internet, which may well be the wave of the future in terms of speed, flexibility, and low cost. Another possibility is the use of easily collected biomarkers of chronic diseases that may track decline in functionality over the life course to supplement ADL and IADL measures that capture changes in functionality across thresholds.

Another approach comes from the Patient-Reported Outcomes Measurement Information System (PROMIS) initiative at the National Institutes of Health, an effort to improve assessment of self-reported symptoms and health-related quality-of-life outcomes using contemporary outcome measurement methodology (Hambleton, 2005). PROMIS is developing and testing a large bank of questionnaire items, some of which may be appropriate for population surveys.

NIA has cofunded and assisted a wide array of surveys across the world, somewhat comparable to the Health and Retirement Study, to learn about the prevalence and the trends of disability in those countries. Cross-cultural and cross-national comparability of measures and data becomes a difficult issue.

Vignettes, as used in the World Health Survey of 2000–2001, is a promising approach to obtain comparability in self-reports of disability in population surveys when a lot of subjectivity is involved across important cross-national population groups. However, this method needs further research; how well it works and in what circumstances has not yet been established.

Given this history and recent developments, a reassessment is timely; NIA asked the Committee on National Statistics (CNSTAT), in collaboration with the Committee on Population (CPOP), to conduct a public workshop on the potential of using methodological advances to improve the measurement of physical and cognitive limitations in population surveys of older adults. The workshop was to focus on disability measures that are suitable for population surveys and that are most relevant for monitoring trends and policy-oriented research and evaluation.

LACK OF CONSISTENCY IN DEFINITION OF DISABILITY

An estimated 40 to 50 million Americans currently live with disability (Institute of Medicine, 2007). Regardless of whether the prevalance of disability continues to decline or increases among the older population, the number of people with disability is going to increase substantially as the

U.S. population ages over the next several decades. The estimates of disability derived from population surveys are therefore increasingly important in order to monitor trends and to track societal progress in the coming years in addressing late-life disability in the population.

The lack of consistency in the concept and definition of disability, and anaysis of trends in the prevalence of disability in the population, has been a real barrier to moving forward in the area of measurement of disability. Because of different views about what is meant by the term disability, over the years researchers have used a variety of measures. Defining disability has occupied the attention of many individuals and groups in the United States for many years. As Jette and Badley (2002, pp. 183–184) explain

> The field of disability research is in need of uniform concepts and a common language to guide scholarly discussion, to advance theoretical work on the disablement process, to facilitate future survey and epidemiological research, and to enhance understanding of disability on the part of professionals as well as the general public. A commonly understood language can also influence the development of public policy. . . . The current lack of a uniform language and commonly understood definition of the concept of "disability" is a serious obstacle to all these endeavors. . . . A common understanding of the term "disability" is an essential first step . . . and is the foundation for a fruitful discussion of improving survey research in the general area of disability. . . . Understanding the source of contemporary conceptual confusion requires a review of the major disability frameworks found in the literature. The goal of bringing together the several different schools of thought on disability and the disablement process remains elusive. Achieving a commonly accepted conceptual language is one of the primary challenges facing the field of disability research. . . .

How disability is defined is profoundly important before grappling with the measurement challenges of how to improve measures of disability, not only within a survey but also comparable across surveys both in the United States and in other countries. If the goal is to devise strategies on how best to "go beyond ADLs and IADLs," the starting point is defining "disability."

The lack of a common definition that can be considered the gold standard shows up throughout the field. The 2003 update of the 1995 paper prepared for the National Institute on Disability and Rehabilitation Research of the U.S. Department of Education shows about 67 federal statutes in which disability is defined (Cherry Engineering Support Services, Inc., 2003). There is overlap among definitions, some statutes have more than one definition, and others refer to definitions contained in other legislation. Broadly, of the 67 statutes or programs that define disability, 35 have self-contained definitions, 26 use definitions from another statute or section, and 6 use definitions from more than one statute or section (Cherry Engineeering Support Services, Inc., 2003).

How disability is defined will, explicitly or implicitly, set the goals for how society chooses to deal with disability. Different definitions will suggest different kinds of solutions and indicate different targets for intervention and action. It is important to reach some clarification on what is meant by the term "disability" (Iezzoni and Freedman, 2008).

HISTORICAL VIEWS OF DISABILITY

To understand the source of contemporary conceptual confusion calls for a review of the major disability frameworks found in the literature. Since the 1960s, survey measures of disability have been rooted in ADLs and IADLs. One of the early large-scale studies was the Established Populations for Epidemiologic Studies of the Elderly (EPESE), which began in the 1980s (see Cornoni-Huntley et al., 1991). EPESE approached physical disability from an operational point of view, focusing on limitation in mobility, ADLs, and IADLs. It used what have become classic batteries, such as a modified Katz ADL scale (dependence in ADLs: walk, bathe, groom, dress, eat, get from bed to a chair, use a toilet) (see Katz and Akpom, 1976), Nagi's functional limitation items from a survey conducted in early 1970s (difficulty to push large objects, stop, crouch or kneel; reach above shoulder level; write or finger or handle small objects) (see Nagi, 1976), and Rosow and Breslau functional health scale (1966) (inability to perform heavy housework, climb stairs, walk a half mile)—all of which focused on very specific ADL- or IADL-type functions. The NLTCS has a heavy emphasis on ADL- and IADL-type metrics with 27 measures of disability that include the Katz ADL index, IADL items from Lawton and Brody (1969), and Nagi's functional limitation items. It also includes some items on vision impairment.

Historically, in survey approaches to assessing disability the definition reflected a medical approach. In the medical view, disability emerges from specific diseases or pathologies. Disability is as an attribute of the person: it is a problem that the person has that is caused by various diseases, trauma, or other health conditions. The management of disability focuses on preventing the condition, curing it, or helping individuals adjust, or change, their behavior to better adapt to it. For example, the Social Security Administration (2003) defines disability as one's inability to engage in substantial gainful activity and this inability is by reason of any medically determinable physical or mental impairment directly tied to an underlying condition. The American Medical Association in 2001 defined disability as an alteration of an individual's capacity to meet personal, social, or occupational demands because of an impairment (Cocchiarella and Andersson, 2001). Disability is directly tied to underlying conditions.

A striking contrast to a medical view is a more social view of disability. This view has not been widely discussed in the aging literature. However, it is prevalent in the rehabilitation literature and the rehabilitation field, in which disability is not seen as an attribute of the individual; rather, disability is viewed primarily as a socially created issue. The crux of this definition is whether or not an individual is or can be fully integrated into roles in society. The inability to be fully integrated is what is defined as disability. For example, Disabled Peoples' International[1] defines disability from the social perspective as the loss or limitation of opportunities to take part in the normal life of the community on an equal level with others due to physical or social barriers.

In the social view, disability is the product of the physical, organizational, and attitudinal barriers in society, which lead to discrimination. It is very different from the medical view of disability. Perhaps the key difference is the assumption that disability is a socially created problem; it is not an attribute of the individual. The defining issue is attitudinal or ideological and requires social or environmental change. Disability is viewed as a human rights issue, not a medical issue. This view is illustrated in the independent living movement and in deinstitutionalization efforts. It is also reflected in the Americans with Disabilities Act.

There have been several efforts to develop a synthesis of both the medical and the social views. For example, there is Nagi's work (1976) and models from the Institute of Medicine (1991, 1997), which defined disability as a phenomenon that results from the interaction between health conditions and contextual factors. They tried to take into account both the medical and social issues and made the argument that disability results from the interaction of both factors.

More recently, the World Health Organization ([WHO] 2001) developed a coherent synthesis view of what constitutes disability. The International Classification of Functioning, Disability and Health (ICF) defines disability as an umbrella term that encompasses differing components, referred to as impairments, limitations in one's activity, and restrictions in one's ability to participate in social roles. Unfortunately, this is not a precise use of the term disability. The more precise terms in the ICF are the impairments, the limitations in activity, and restrictions in participation. The WHO's umbrella term for this collection of concepts is disability.

At the level of body functions and body structures, the WHO document refers to such conditions as hemiparesis (muscle weakness affecting only one side of the body) secondary to restricted blood flow, for example, to the brain—restrictions occurring at the level of organs or body systems. It

[1]Disabled Peoples' International is a network of national organizations or assemblies of disabled people, established to promote their human rights through full participation, equalization of opportunity, and development.

also discusses limitations in activities or restrictions in participation, such as, functionally, an inability to walk, severe difficulty in dressing oneself, or difficulty in preparing meals. These limitations are similar to ADLs and IADLs, but they also include much broader social role behaviors, particularly under the concept of participation. Then very explicitly, the model tries to incorporate factors of the environment, such as social, attitudinal, and physical barriers and facilitators that affect either the body functions and structures or activities and participation. That is how the WHO has tried to pull together a synthesis of some of the different views of what constitutes disability.

WORKSHOP PLAN AND REPORT OVERVIEW

The purpose of the workshop, as noted earlier, was to draw on the most current understanding of methodological advances to improve the measurement of physical and cognitive disabilities in population surveys of older adults, focusing on disability measures that are suitable for population surveys and that are most relevant for monitoring of trends and policy-oriented research and evaluation. To respond to these issues, CNSTAT, in collaboration with CPOP, appointed an expert workshop steering committee for the task. The committee was charged to plan the workshop, decide on commissioning technical background paper(s) as needed, develop the agenda and identify researchers for the presentations, and convene and serve as moderators for the workshop.

The steering committee believed that a focused discussion among a wide range of disability researchers and survey methodologists could identify a framework for a long-term research agenda in this area for NIA and others in the field. To contribute to that framework, the committee commissioned a background paper *Population Survey Measures of Functioning: Strengths and Weaknesses* by Barbara Altman, which is included as Appendix A in this volume. The workshop drew people from a wide variety of disciplines and perspectives from federal agencies, academia, nongovernmental organizations, and the public. The workshop agenda and presenters and biographical sketches of steering committee members are shown in Appendixes B and C, respectively.

The workshop was structured to combine formal presentations and discussion among the participants. This report is a summary of the presentations and the group discussions flowing from the presentations during the sessions outlined in the agenda. The next chapter opens with an overview of the technical background paper, then discusses the challenges on improving current measurement of late-life function and disability. Chapter 3 focuses on potential methods for refining or augmenting current measures. Chapter 4 addresses issues related to improving the validity of cross-population comparisons, and Chapter 5 addresses the need to measure functioning and disability in context. The final chapter focuses discussion on the future in terms of suggesting research areas to advance the current efforts from the perspective of the participants attending the workshop.

This report is a summary of what transpired at the workshop and therefore limited to the views and opinions of those participating in the workshop. It reflects the concerns and areas of expertise of the workshop participants. As such, it does not provide a comprehensive review of the research and current status of survey measurement of disability.

2

Challenges to Improving Measurement of Late-Life Functioning and Disability

This chapter summarizes the first workshop session designed to explore the challenges to improving current measurement of late-life functioning and disability in population surveys. The session opened with an overview of the background paper commissioned for the workshop (see Appendix A). It was followed by presentations on four topics:

1. Developing questions in surveys to identify people early in the disablement process
2. Enhancing the ascertainment of disability
3. Self-responses versus proxy responses in surveys
4. Expanding modes of survey administration

POPULATION SURVEY MEASURES OF FUNCTIONING: STRENGTHS AND WEAKNESSES

Barbara Altman (disability statistics consultant) presented a brief overview of her background paper (Appendix A). She noted that including measures of disability in population surveys involves many disciplinary approaches and many research agendas. As Nagi pointed out in 1991, the history of the involvement of many disciplines in the development of theory, policy, and programs to address disability issues—medicine, education, social work, psychology, sociology, vocational counseling, occupational and physical therapy, and others—sets the stage for attempts at conceptual distinctions to delineate measures of the disability process. There has been a tremendous richness in the work and the number of disciplines involved in trying to understand disability, trying to develop models of disability, and trying to identify measures of it. Yet this variety in approaches, language (and jargon), orientation, and focus complicates efforts at measurement and sometimes confuses interpretation of the results of the measurement.

Altman stated three objectives of her paper:

1. To examine the disability conceptualization transition to measurement within the general theory of disability and to compare it across areas of application
2. To define the sources and types of measurement that have been developed for the various theoretical concepts and examine what measures are available and how well they represent the concepts
3. To reintroduce the important contribution of the social and environmental context, not only to the conceptualization of disability, but also to its measurement

Altman then gave an overview of the measurement elements that are necessary to fully understand disability, reveal the strengths and weaknesses of what we have, and identify the gaps in measurement that exist.

Transition from Concepts to Measures

Multiple theoretical models provide the conceptual basis for understanding the disablement process. The Nagi model (1965), developed by a sociologist, is one of the earliest and most widely known coherent organizations of the conceptual components and their relationships. It was revisited and expanded in 1991 by Nagi and also elaborated by Verbrugge and Jette (1994).

Subsequently, conceptual elements and relationships have been expanded with models from the Institute of Medicine (1991, 1997) and the International Classification of Functioning, Disability and Health (ICF) model of the World Health Organization (2001), but their elaborations do not really take new directions. They do make the components more understandable to a wider audience, provide a standardization of the language, and make things more accessible to the people who are using the models. The ICF model also provides an accompanying classification scheme that is a listing of domains for consideration when one is operationalizing a measure. It has been a very useful tool. While each of these succeeding models has made contributions, the original model is still very visible.

On the basis of these models, major conceptual elements that make up the experience of disability and need to be measured in population surveys include (the background paper and this discussion focus only on those concepts that are starred)

- pathology or impairment,*
- personal factors,
- functioning of the whole person,
- actions or activities,*
- participation or disability,*
- environment,* and
- quality of life.

Personal factors, at least in terms of demographic characteristics, are commonly measured in population surveys, although some additions could be made for the purposes of studying disability. Quality of life reflects a variety of conceptual composites that have been constructed differently in different research situations.

The transition from theory to measurement involves several steps and several areas of decision making that are not always thought through when the process is happening. Yet this transition is the point at which the theoretical elements are converted to operational choices, either specific characteristics of, or observations about, individual respondents. Altman noted that the process in effect creates the breadth or limitations of the data to reflect the concept being addressed and, as such, requires conscious consideration and forethought.

The primary conceptual components of disability models have become familiar to people at the abstract and theoretical levels. However, translation of the concepts into concrete (i.e., everyday reality) measurement in a population survey involves decision points, and with each decision point the initial basic concept is narrowed. Because of the limitations associated with population surveys—time, space, and cost—it is hard to include all the conceptual elements identified in a full theoretical model of disability.

Briefly, the transition from theory to measurement process involves

- identifying the purpose of data collection,
- identifying the appropriate conceptual component related to the purpose of the data collection,
- operationalizing the theoretical concepts in real-world terms—deciding what actual behavior or characteristic will represent the larger concept, and
- locating the unit of analysis and the type or level of measurement.

Purposes for Measurement

The Washington Group on Disability and Statistics, an international organization that is seeking to develop comparable measures of disability internationally, has identified three major purposes for data collection in population surveys.

The first purpose is to identify trends in prevalence rates of impairments, social limitations, or levels of participation. Trends can be developed of almost anything that is measured so that construction of the measure creates the bounds of the population examined. For example, if ADLs and IADLs are measured, one only identifies trends in the prevalence of limitations in ADLs or IADLs. One cannot assume that those identified by the measure also represent all persons with physical or sensory limitations. While it is likely that all the persons who indicate that they have ADL and IADL limitations also have physical, sensory, emotional, or cognitive limitations, they do not represent the total population with all those kinds of limitations. Similarly, if the measures focus on physical or sensory limitations, the resulting trend data document the change over time in those limitations. Such measures are also likely to include most people with ADLs or IADLS, but because the measures are of physical or sensory functioning, they may cover a much larger population. People with ADL or IADL limitations can only be assumed to be a smaller and less identifiable portion of the population than has been defined. The purpose of trend data is simply to monitor the changes in prevalence of a certain conceptual element of disability over time.

The second purpose of collecting data in a population survey is to demonstrate the provision of service and programmatic needs of a population. Measures of service needs are generally focused on particular types of impairment, functional limitation, or age groupings and usually involve such subpopulations as wheelchair users, people who have problems communicating, or people with spinal cord injuries. Much more detailed information is needed about those subpopulations in order to provide the information required to develop programs or document that programs are working. National (or general) population surveys are not necessarily an appropriate vehicle for that kind of purpose because of an insufficient sample size for the specific problem or because of the number of questions needed to provide the necessary detail. Such information is best obtained in a medical setting or in a special survey.

The third purpose of collecting data in a population survey is to assess the integration in or the equalization of opportunity for a population of concern as compared to the general population. This is a new approach to understanding disability and is related to the social model. It addresses the interactions of people in their communities and calls for a measure that identifies the full range of possible candidates for discrimination—the "at risk" population. It derives from the Americans with Disabilities Act, which guarantees the rights of all people with disabilities, including access to buildings, doctors' offices, stores, jobs, and all other public places without discrimination because of physical, mental, or emotional difficulties. It is very similar to what has been done relative to race, gender, and age discrimination. Disability measurement for this purpose takes on a much broader focus than is found in the other two purposes. It calls for measures that identify a full range of people at risk of discrimination because of their limitations.

The questions in the American Community Survey that went into use in 2008 were specifically developed to satisfy this third purpose of monitoring equalization of opportunity. Respondents are asked about functioning limitations with the assumption that those who have functioning difficulties of any kind are the population who are most at risk for limitations in social functioning because of the social or physical structure around them. The measures are then used as a demographic context to examine differences in access to education or participation in employment in the same way that one would examine differences in access between men and women or among racial and ethnic groups.

Operationalization of Measures

Altman noted that when operationalizing a theoretical concept, such as functioning, one moves from the conceptual definition, which incorporates all possibilities, to the single question or observation that represents one possibility or a small group of them. One has to decide what measure is going to represent functioning.

There are also various levels of complexity of the measurement of disability. Some things are relatively simple, such as whether an impairment or a condition is present. That is a yes or no question, and whether it is a self-report or a doctor's diagnosis is used, it gives a good idea of a person's general health condition. However, the current models of disability reflect a hierarchy or an increasing complexity of the components associated with the disability process. In addition to the representation of the presence of an impairment or an impairing condition, there are at least four levels of measurement reflected in disability conceptual models: basic action, specific task, organized activity, and role participation. Each of these represents a more complex level of action or activity. The areas of activities and participation identified in the ICF model, known as domains, include all four levels of complexity in a particular area. An example would be the area (or domain) of mobility that incorporates activities, such as changing body positions, moving and handling objects, walking or moving to different locations, and moving around using transportation.

The levels of complexity of conceptual components influence how a measure is constructed. Generally, questions about basic actions or movement or use of the body or mind represent the simplest level of action or activities. A more complicated level of action or activity that goes beyond coordination of bodily movement is represented by the specific task that an individual is motivated to accomplish and requires a combination of physical movements, sensory perception, intellectual activity, and possibly the use of assistive devices in order to complete the task.

The next level involves combining basic actions and multiple tasks to accomplish what may be considered a behavioral element of an ongoing role: For example, using a motor vehicle is an organized activity that is useful in many roles. As a parent, taking children to school fulfills one of many parental roles, but as a bus driver the ability to drive a motor vehicle is one of the central and necessary elements of the job. Finally, participation represents the accomplishment of enough organized activities to satisfy minimum role requirements to be considered integrated in a specific role.

Measures in Population Surveys[1]

Almost all population surveys with disability measures include impairment measures. Measurement of functioning is frequently associated with questions developed by Nagi to measure physical functioning in the employment context; such measures represent the whole person's ability to accomplish basic actions, such as walking, seeing, and communicating. In contrast, measures of behavior or functioning in social roles are much more complicated and complex. Disability is experienced when the person with the functioning limitation interacts with the cultural expectations or the physical environment. There are far fewer measures of this type in surveys, and it is probably the direction that measurement work should take to understand the full effects of functional limitations.

In summary, Altman said, there is a rich set of survey measures on the task level that represent dependence in basic task activities that are necessary for maintaining independence. There are also some good representations of physical functioning, although not all areas of physical functioning are well represented. Other types of functioning are not well represented. Mental health or intellectual functioning is not well represented, although there are some intellectual tests that are included in surveys of older people. Large gaps exist in measuring role participation, as well as both the physical and social environmental contexts in which all action takes place. There is a great lack of standardization of core measures that permit a continued search for uniform concepts and a common language. Without standardized core measures, one cannot accumulate knowledge in a way that is

---

[1]The background paper (see Appendix A) has an extensive examination of these measures in tabular form for the most frequently used population surveys with disability measures.

useful because each measure represents a somewhat different segment of the population with disabilities.

DEVELOPING QUESTIONS TO IDENTIFY PEOPLE EARLY IN THE DISABLEMENT PROCESS

Linda Fried (Mailman School of Public Health, Columbia University) began her presentation with a question: How can one identify people early in the disablement process? In other words, is there life before ADL and IADL limitations—looking at disease, predicting impairments, predicting functional limitations, predicting disability? The natural history of disability indicates that at least half of disability in older adults is chronic and progressive (Ferrucci et al., 1996). Catastrophic progressive mobility difficulties, and other factors, predict difficulty or independence in ADLs and IADLs (Harris et al., 1989), and having difficulty in a task predicts dependency (Gill et al., 1998). Given this chronic, progressive course, the key to prevention and compression of morbidity is early ascertainment.

A number of survey measures have been developed over the past several years to ascertain individuals at an early stage of disablement, those who are likely to be most amenable to interventions, with the goal of effective targeting as a basis for intervention. They include

- a focus on mobility, as well as social roles and compromise of them as disability progresses;
- "life-space diameter" and its constriction over time in terms of the activities in which people engage and the geographic perimeter of their lives, as markers of people who are on a disablement pathway;
- disability in more demanding tasks, as a predictor of disability in less demanding tasks;
- fatigue and tiredness as early indicators of functional decline and predictor of incipient disability; and
- preclinical disability, including survey or objective measures, as well as mixed measures combining both survey and objective performance-based measures, such as screening nomograms for preclinical disability and measures of frailty, both of which predict incident disability.

To illustrate the disablement process, Fried offered a story of a 75-year-old woman with arthritis of the knees and fear of falling: About 10 years ago, she started having difficulty walking a half mile and since has stopped being able to do that. Then, 2–3 years ago, she started having difficulty climbing stairs and would clutch the handrail. About 2 years ago, she stopped being able to do heavy housework and cut back on light housework, and about a year ago she stopped being able to carry groceries from the store. She said that difficulty in each of these tasks was due to the same reasons: arthritis of her knees and her fear of falling (Fried and Herdman, 1992). This story provides a sense of what the disablement pathway might feel like to an individual.

Much interest exists in whether self-reported task difficulty in more demanding tasks (such as this woman was describing) can, itself, identify people early in the disablement process and at risk of progression to difficulty in less demanding tasks. It raises the question of whether there is a hierarchy in a group of tasks, such as mobility. This question raises a further question: Can optional participation in higher level, complex tasks be used for ascertainment of those at risk of progression to disability? Reuben and colleagues (1990) proposed looking at advanced ADLs, for example, how much exercise people are engaging in as a predictor of their level of activity.

Self-reported task difficulty in more demanding tasks can, itself, identify people early in a disablement process and possibly presage incipient disease. For example, unpublished data from the Cardiovascular Health Study (Fried et al., 1991a)—a prospective, observational cohort study of older adults in four U.S. communities—suggest that onset of reported difficulty in these tasks might have the added interesting feature of predicting incipient disease events. The study looked at physical function before and after cardiovascular events, and tasks were divided into three categories as being more or less demanding in terms of exercise tolerance, but all requiring some mobility. One category involved more demanding tasks, such as walking a half mile and climbing steps; another category included medium-demand activities, such as shopping and preparing meals; and a third category included low-demand activities, such as walking around the home or getting out of a bed or chair. Looking at the months before or after the onset of a cardiovascular event—coronary heart disease, congestive heart failure, or stroke—a precipitous decline occurred in the proportion of the population without difficulty in demanding tasks, although there was little change in the proportion with regard to low- or medium-demand tasks before the event itself. This suggests that the onset of difficulty in high-energy-demand tasks indicates a decline in physiological status before the cardiovascular event itself. There is a need to characterize what that is: Is there a preclinical stage of change in physiological function marked by onset of some disability? Would identification of that stage help find the people who are really at very high risk of progression of disability? Those people would be a particularly desirable target for intervention and likely to be much more amenable to improvement than after they have become severely disabled.

Going back to the example of the woman with arthritis, Fried said that there were a number of less demanding mobility tasks that the woman had no difficulty performing, but over the same 10-year period she had started to change the way she did them (Fried et al., 1991b). This phenomenon suggests that there is a progressive process even among people who have not reported difficulty following a hierarchy of mobility tasks. At early stages, people may be able to successfully compensate for the effects of disease and maintain their function without difficulty.

Fried stated that many years ago she and her colleagues hypothesized that there might be observable preclinical changes in function that could identify an early stage of disablement and that they might be able to ascertain such changes through survey methods. This intermediate stage of function between high function and difficulty in a task would be measured by self-reports of whether people modify task performance or its frequency as a result of underlying changes in health, even though they reported no difficulty. What would really be measured is adaptation to physical limitations in order to preserve task performance.

The Women's Health and Aging Study (Fried et al., 2000) found that, among the two-thirds least disabled older women living in the community who had no difficulty at the beginning of the study, those who were reporting task modification were at fourfold higher risk of developing difficulty walking a half mile over 18 months (adjusting for a number of factors). Walking speed was also an independent predictor, with a twofold increased risk for lower walking speeds in these models. Interestingly, strength and other covariates in these higher functioning women were not predictors of subsequent changes for them.

Another example is a series of studies by Douglas Miller and Fred Wolinsky in St. Louis (see Miller et al., 2005), which looked at inner-city African Americans in comparison with older African Americans living in suburbia and whites living in suburbia. In the first study, the authors found that inner-city African Americans who were 65 and older were already so substantially disabled that preclinical disability could not even be ascertained. They found that disability was occurring about 10 years earlier in inner-city African Americans than in suburban African Americans or whites. They then did another study to look at middle-aged African Americans (ages 49–65) living in inner cities and found that 60 percent already had one or more disabilities and that about 33 percent already were reporting preclinical disability in mobility tasks.

They also found that preclinical status in terms of mobility in these middle-aged African Americans living in the inner city predicted a four-and-a-half-fold increased risk of onset of difficulty in walking a half mile, with preclinical disability again defined as self-report of modifications in the way people went about doing mobility tasks but without difficulty. An interesting finding from this population is that there were no physical performance tests that were predictive of incident mobility difficulty, while the self-report survey methods were highly predictive in the same model. These are somewhat different findings than found in the Women's Health and Aging Study of an older group of women. Wolinsky and colleagues (2007) concluded that the preclinical disability survey measure they used was a highly effective early warning system and a target for prevention.

Other work in the Women's Health and Aging Study suggests that the way people modify mobility tasks has a lot of variation, but in general suggests a hierarchy of compensation use, going from the very intrinsic ones (such as change in pace or biomechanics) that do not threaten people's own perception of whether or not they are having difficulty, to highly extrinsic ones for which the modifications are more evident. Basically, people report doing the task more slowly, changing their body position, or doing the activity less frequently; then they cut out parts of the activity that they would normally do in a day; and then they start using assistive devices and human assistance.

This work offers one perspective on using surveys to identify people with earlier changes in function in a way that can be used for targeting for prevention of disability.

ENHANCING THE ASCERTAINMENT OF DISABILITY

In his presentation, Thomas Gill (Yale University School of Medicine) summarized some of the results from the Precipitating Events Project, a prospective cohort study of 754 initially nondisabled persons aged 70 years and older living in the community that he has been leading for the past decade (for further information, see Gill et al., 2002; Gill and Gahbauer, 2008; Gill et al., in press). Monthly telephone interviews have been conducted for up to 130 months assessing disability in ADLs, IADLs, and mobility. Gill focused on the four essential ADLs: bathing, dressing, walking, and transferring.

A standard strategy for ascertaining the occurrence of disability does not currently exist. In most longitudinal studies, an incident episode of disability is noted when a nondisabled person reports disability "at the present time" during a subsequent follow-up assessment. Yet it has been shown previously that incident episodes of disability are often not ascertained by longitudinal studies with assessment intervals longer than 6 months. Using the traditional strategies, for individuals nondisabled at baseline who are assessed again 1 year later, the incidence rate of disability might be about 2.5 percent. Yet by evaluating them every month, as was done in this study, at the 12-month mark the disability rate was 10 percent. This finding indicates a substantial underestimate of the incidence of disability in studies with infrequent assessments.

The question is: What is driving these underestimates? Analysis of data for 24 months showed that for each of the three risk groups—low, intermediate, and high—the difference between the cumulative disability rates increased progressively as the length of the assessment interval increased. Although these differences in rates were attributable almost exclusively to recovery from disability from 1 month to the next in the first 6 months, they were due increasingly to deaths and some losses to follow-up over the next 18 months, especially among participants in the high-risk group.

To evaluate whether the ascertainment of disability could be improved in longitudinal studies, the researchers added several questions that had not been included in the prior assessments to the comprehensive assessment at 72 months. In addition to asking about disability at the present time, for each of the four essential ADLs, participants who did not need help from another person "at the present time" were asked to recall whether they needed help from another person to complete the relevant task "at any time" during the past 1 month, 3 months, 6 months, and 12 months, respectively. Focusing on the 12-month results, up to one-half of the incident disability episodes, which would otherwise have been missed by asking only about disability at the present time, could have been ascertained if nondisabled persons had also been asked to recall whether they had had a disability at any time since the prior assessment.

At this new 72-month baseline, 370 people were not disabled in their basic ADLs when first surveyed. One year later, 14.2 percent (53 people) said they were disabled at the present time. That is the standard incidence rate for disability that would be determined from traditional surveys. However, by asking those individuals who were not disabled at the present time the additional question, "At any time during the past year have you been disabled?" almost 13 percent more (48 people) responded in the affirmative, for a total incident rate of 27 percent—almost double what would be found using the traditional approach.

Gill said that the next question was how to determine if these reports of disability at any time over some period were accurate. People who were not currently disabled were evaluated in two groups: those who said they had not had any disability at any time during the past year and those who said they had. The two groups were followed forward for another 18 months, with the hypothesis that people who reported having a disability or having had one at any time over the previous year would have worse outcomes, and that is what was found. Specifically, the additional disability episodes ascertained only by the person's recall predicted high risk for the subsequent development of chronic disability—a major determinant for the use of long-term care services—even after accounting for potential confounders. This finding provides some validation that the reports of disability at any time are valid.

Despite this potential advance in the assessment of disability, a large proportion of older persons do not recall episodes of disability that in fact occurred during the prior year. An effort to identify the factors that are associated with accurate recall of prior disability found that participants' education was the strongest predictor: more highly educated individuals were more likely to accurately recall having had a disability over a preceding 12 months. Cognition was also associated with accurate recall, but not statistically so. Although education could be a proxy for cognitive status, the effect of education was not attenuated in the multivariate analysis, which included cognition as a covariate. The validation effort also found that disability-specific factors had some relationship to accurate recall. Thus people who had a disability episode more recently, say within the past 3 months, were more likely to recall having it than those with less recent episodes. Those who had at least one episode of severe disability, as well as people who had had a severe episode (defined as having disability in three or more ADLs), were more likely to recall it than others. If a disability persisted for more than 1 month, the likelihood of recall was higher.

In summing up the results, Gill offered ways to enhance the ascertainment of disability. If an individual is not disabled at the present time, ask whether he or she has had a disability episode at any time since the prior assessment. If the response is no, probe further, using a standard protocol focusing on major illnesses or injuries that have occurred since the prior assessment. Special attention may be warranted for people with low levels of education and perhaps those who are cognitively impaired. Lastly, another way to possibly enhance the ascertainment of disability would be to adopt a calendar approach, which has been used successfully to ascertain falls. Some variation of this approach could be implemented for the ascertainment of the incidence of disability.

SELF VERSUS PROXY RESPONSES IN SURVEYS

Jay Magaziner's (School of Medicine, University of Maryland, Baltimore) presentation dealt with the use of proxies to obtain information on health and functioning of older persons in population surveys, describing some of the issues, suggesting some practical strategies for using data from proxies, and identifying areas for additional study. He noted that there have not been any major breakthroughs in the use of proxies in the past several years, and so a lot of the information comes from work that was done some time ago.

The significance of the problem is obvious. There is a substantial increase in the number of older people in the population, and this number is projected to increase. There is an increased need to conduct clinical and population research on this group; the omission of persons who cannot respond for themselves limits generalizability in research. There is a lot of effort toward improving measurement of disability, but a key question is how to get information about disability from people who cannot report for themselves.

Proxies are used to obtain information about people who cannot respond for themselves, will not respond for themselves, or are difficult to locate initially or for follow-up. At times, proxies may be used to obtain information in a less costly way. However, if proxies are used, how would one factor the information and what would be its real utility?

The extent of the problem of nonrespondents varies depending on the group of interest. Among people aged 65 years and older, about 5–10 percent of community-dwelling people are unable to provide reliable information for themselves because of cognitive limitations. As many as 40 percent of people who are hospitalized are unable to provide information for themselves. For nursing home residents, this number is well over 50 percent. Depending on the group of interest, one is dealing with a fairly large problem. In addition to people who cannot respond because of cognitive limitations, there are people who cannot respond for other health reasons, people who refuse, and people who cannot be located. Thus, there is a fairly sizable problem of missing information if it cannot be obtained from some other source. Areas of measurement for which proxies may be needed include measures of health status, including information on reported diagnoses and symptoms, and a variety of areas of functioning—physical, instrumental, affective, cognitive, social, and economic. There may be other areas, but these are some that have been examined.

The major issues related to subject and proxy agreement are no different from the kinds of issues faced in any kind of scientific measurement. These are issues of precision and bias. The level of agreement between subjects and proxies is really a function of the precision and the bias (bias refers to discrepancy and not whether one is right or wrong). Most studies have focused on agreement, but essentially, because of the nature of agreement being a composite of precision and bias, less has been done on bias. Magaziner noted that researchers need to pay more attention to the magnitude and direction of bias. Precision is important, but when using proxies, one must consider bias. Agreement and bias are both functions of the question asked, characteristics of the subject, characteristics of the proxy, and characteristics of the context and culture.

Magaziner next highlighted some of the findings of studies of patient and proxy responses. In a study of community-dwelling women aged 65 years and older, those who had been hospitalized for hip fracture were asked about their ADLs, walking, and how they were before they were hospitalized. When both subjects and proxies were asked about a simple task, such as walking ability, the measure of agreement was fairly good, with 10 percent of the patients reporting they were unable to perform a walking task or needed a lot of assistance with it, and proxies reporting about 10.8 percent (Magaziner et al., 1996). The level of agreement declines in an ordered manner as one moves from walking to bathing, shopping, preparing meals, dressing, handling money, and grooming. The level of agreement for handling money was not very good, possibly because of the complexity of the question. The question itself may not be tapping into the same thing for a self-report and a proxy: 15 percent of subjects reported that they could not handle money on their own, while 20 percent of proxies said they could not. Whenever there is a bias, it tends to be in the direction of more disability reported by proxies.

With regard to affective status, the subjects were asked about depressive symptoms, using a Center for Epidemiologic Studies Depression Scale type of measure on them. Then the proxies were asked how they thought the person would respond to those particular questions. The same was done with cognitive status, using the Mini Mental State Exam. The bias was quite small, but in this community survey there was a negative bias. The proxies underreported both depressive symptoms and cognitive status compared to the subjects (Bassett et al., 1990). Similar analysis was done with data on a post-hip fracture group; for which proxies overreported depressive symptoms but clealy underreported cognitive problems; that is, proxies said that the person actually performed better than was shown in a test of cognition.

For people with ADL and IADL limitations due to chronic conditions, there is a tendency for proxies to overreport disability (Magaziner et al., 1996). For physical symptoms experienced within the past month, there is no consistent pattern. Many of these symptoms are private symptoms. They are not things that a proxy would know easily, which may result in a lack of agreement.

Some characteristics of the proxy make a difference. For example, female proxies tend to report more disability than male proxies when their responses are compared with the subjects themselves. Proxies who live with the subjects report more disability than those who do not. Those who assist subjects report more disability, and those who claim to have good knowledge of the subject generally report more disability than those who do not.

To summarize, proxies can provide answers that agree with subject reports for objective, observable items, such as walking, and chronic disease states. Proxies are poor reporters of private unobservable items, such as the use of a urinary catheter or symptoms. Proxies are poor reporters of complex tasks when the questions are asked in a global manner, such as handling money.

When there is disagreement, proxies generally report higher levels of disability than subjects report for themselves, with the notable exception of cognitive function. Female proxies, those living with subjects, and those providing care report more disability than subjects report for themselves. Agreement and bias are functions of the question, subject characteristics, and proxy characteristics.

There is some practical advice about using proxies that one can take away from this work: Develop more objective questions that do not call for judgments by proxies. Conduct pilot studies for questions to be used and proxies to be encountered in the population under study, and try to understand how the proxy would perform in that particular study and then use that information for interpretation of results. Evaluate agreement and bias. Consider using only proxies, which might be useful if a large percentage of dropouts is expected: Why introduce another level of bias if one can get all the information consistently over time from proxies?

Further research is needed to evaluate proxy data for those who cannot respond for themselves. Most of the research to date is based on subjects who can respond for themselves. It is important to develop and test better questions and determine whether data adjustments can be made from knowledge about questions, subjects, and proxies. Evaluate the effects of substituting proxy data on parameter estimates; evaluate the effect of using only proxies, especially when bias is significant, and evaluate the effects of using information from multiple sources in the same analysis to arrive at assessment of functional status.

In conclusion, proxies can be used with a reasonable degree of reliability for some questions. More research is needed on the use of proxies for measuring functional status in those who cannot provide information for themselves. Proxies must frequently be used in place of subjects in studies of older persons, until some good methods are developed for obtaining information in a reliable way about those who cannot provide it for themselves. Scientists have an obligation to report on their use of proxies and describe the possible effects that they can have on study results.

EXPANDING MODES OF SURVEY ADMINISTRATION

Arie Kapteyn (Labor and Population, RAND) addressed the use of the Internet for survey administration as the focus of most of the innovation at this time, but some of the issues he raised are also relevant to other modes. He focused on Internet interviewing in general and the Internet and the elderly because disability clearly is most prevalent among the elderly, and it is also the group for which Internet use is still more problematic than for other age groups. He also discussed new technologies and some perspectives on what is coming next.

Internet penetration in the United States is probably about 75 percent. In Europe the Internet penetration is probably about 50 percent, with large variations among countries. In the Netherlands it is about 90 percent, probably even higher. Scandinavia is also very high. In Southern Europe it is much lower: In Greece penetration is only 35 percent. In countries with low Internet penetration, using the Internet as the only survey mode would lead to coverage error. Yet other modes, such as the telephone, also have problems; telephone interviews that use only land lines increasingly suffer from the same problem of coverage, and there are also problems because of answering machines that screen calls.

Internet coverage is directly related to age. In a study using the Internet mode, researchers found that in 2002 almost half of respondents under 60 years of age had Internet access. This number declined quickly until only about 10 percent of the respondents 76 years and older had Internet access (Couper et al., 2007).

Data from a new panel set up in the Netherlands (the Longitudinal Internet Studies for the Social Sciences) provide some information on how representative a study can be using the Internet for survey administration. For this Internet panel, respondents received broadband Internet access if they did not have it yet. In collaboration with Statistics Netherlands, the researchers used population registers as a sampling frame. Kapteyn remarked that one of the great things about northern Europe is that there are population registers, which make nice sampling frames. The baseline response rate of this panel was 50 percent, which for a panel in the Netherlands is quite good.

One of the things that people always talk about is the mode effect. What is different between the Internet and other modes? Internet and written interviews are similar, and computer-assisted telephone interview (CATI) and computer-assisted personal interview (CAPI) are also quite similar. Essentially, the distinction really is if there is an interviewer.

Some features of the Internet that make it attractive are speed and cost-effectiveness, especially for panels. Once people are in the sample, questions can be asked at any time of the day and any time of the week. However, it is also dangerous in the sense that anyone can do Internet surveys. Arduous tasks can be broken up into modest-sized bits: For example, if you do surveys of no longer than 30 minutes over a couple of weekends, you can amass a lot of information. A total of 5 weekends of 30 minutes would yield 2.5 hours of interviews. Other attractive features are quick turnaround, feedback, flexibility, and high-frequency and event-related interviewing (e.g., following the onset of disability or some illness). In terms of automation, the Internet mode has all the advantages of CATI and CAPI. One use of this approach has been the American Life Panel, which since November 2008 has been monitoring via the Internet how households are faring in the financial crisis. Various experiments have been done on elicitation of probabilities and expectations, portfolio choice and presentation of information, a sequence of vignettes in the Netherlands and the United

States, including test of response consistency, and animation. The Health and Retirement Study (HRS) instrument is being migrated to the Internet; it is expected to be completely on the Internet in about a year. It is administered to respondents in chunks. Respondents get a module, and a couple of weeks later they get another one.

Some examples of future possibilities are a heart rate monitor and an actigraph device that measures individual activity level, energy/caloric expenditure, duration and intensity of sustained activity, daily activity profile, limb extremity movements, sleep patterns and night activity, steps taken, and heart rate (in at least some models). The input from these devices can be combined with the Internet. The respondents are asked to wear the device, say for a week, and the measurements can easily be transferred by using wireless technology, or a USB key to transfer data by computer or by mailing the device back. These measurements can be combined with self-reports of activities or stress, time-use data, self-reports of subjective well-being, experience sampling, anchoring vignettes, etc.

One reason for the interest in using these devices is a result of findings from the HRS and its English equivalent, the English Longitudinal Study of Ageing: In questions about physical activity, Americans say they are more physically active than the English, and somehow the English do not believe it. There is currently a proposal to use these devices in the United States and England and find out whether the English are too modest, the Americans are bragging, or something else is going on.

In conclusion, Kapteyn observed that because Internet penetration is related to age, it is likely to grow substantially, even among the elderly, as cohorts age. In addition, the user friendliness of devices is improving quickly. Finally, more attention should be devoted to design of websites intended for the elderly.

DISCUSSION

In the discussion, the topic of proxies and proxy measurement drew the most comments. Other issues of note were use of the Internet for data collection, the role of the home environment for conducting performance tests, and phobias in old age.

Use of Internet for Data Collection

A participant noted an interesting aspect of using the Internet for data collection, as well as any of the research looking at age differences and use of the Internet, namely, that there actually are age differences in sensory perception and ability to physically use a computer interface. Arie Kapteyn was asked if design issues related to each of the cohorts were examined. Although a lot of these concerns will be moot in the next several years as greater familiarity with computers and the Internet moves through the population, design issues are important when the Internet is the mode of data collection, especially for the cohorts in the older ages.

He responded that they had not addressed that issue in their study, but there are people working on website design for all age groups. The Internet is still very much attuned to young people. The smaller devices with a lot of information are really difficult to read. For very old respondents, the first thing needed is that the letters have to be big, and the screen should not be cluttered—it has to be as simple as possible, otherwise people get confused.

Role of the Home Environment for Conducting Performance Tests

Linda Fried was asked if in her studies she and her colleagues had studied the differences that the home environment makes in conducting performance tests with different populations. Lack of space to set up the walking speed course could limit the ability to do the performance tests.

Fried responded that she was not aware of anyone having looked into this issue. In the design of the Women's Health and Aging Study, investigators spent a lot of time on the design of those performance measures, and they were able to do performance-based measures on highly disabled older women in some pretty constrained homes. In the Whitehall Study of British civil servants (Brunner et al., 2009), they took what was then the standard 4-meter walk and if the space was too limited in the home, designed a way to do it in just 3 meters in a very standardized way. They were then able to compare both the 4-meter and 3-meter walks in the same data set.

Phobias

Robert Wallace (University of Iowa) asked Fried if the fear of falling expressed by an individual in her study was due to a phobia about falling or the result of disability. He said that there may be a lot of phobias in old age, which may or may not be warranted, that researchers do not pay much attention to fear of falling, fear of crowds, fear of noise, and fear of going out. Such phobias may be a lot of the reason for a "disability" rather than actual mechanical problems of the disability. There are a lot of phobias without a physical basis that create a fear that may then affect functioning. There is a whole range of other things that also affect behavior and what may be perceived as disability.

Fried responded that in that particular case the person's fear of falling was entirely due to the instability of her knees from osteoarthritis. She was not phobic. However, Fried agreed that there are a lot of phobias without a physical basis to create that fear that may in fact be modifiers. For example, does a person have a reason to get up in the morning? Depression aside, are there activities that one has available that one cares about? Are there places to go? All of those things affect motivation, absent psychiatric illness, which are also huge modifiers of behavior.

Proxies and Proxy Measurement

Robert Hauser (University of Wisconsin-Madison) commented on the importance of the gold standard with respect to proxy measurement. All survey responses are subject to error. If the correlation between self- and proxy reports is high, it means that people are reporting perfectly up to the level of reliability of the instruments. One needs to think about measurement error on both sides and about a gold standard for the value of proxy measurement.

Jay Magaziner responded that the need for a gold standard really gets more to a fundamental question that cuts across all of what has been discussed. That is the purpose of measuring something before one can talk about how best to measure it. If there is a reason behind what one is measuring, whatever it is, then one might be able to approach some kind of a gold standard for that specified purpose. So the key question is what does one really want to know and why? The gold standard is now a mixture of the environment, the social situation, the nature of the items we are asking, and so on.

Fried commented that this issue of a gold standard is something that has many different dimensions. It goes to the issue of both what one wants to understand and also the experience of the individual. There are many contextual factors that shape or modify and exacerbate or minimize that human experience. The human experience of what people are able to do is the core issue. There is as yet no conceptual agreement about measuring disability.

In addition to the discussion of a gold standard for proxy measurement, other issues about proxies drew lively discussion. Participants reiterated several points made in the presentation: One of the reasons for bias in responses in proxy measurement might be characteristics of the proxy, such as gender. Is there similar information on cultural or ethnic differences that might account for different perceptions of disability in the subject? Magaziner said that female proxies tend to report more disability in the subjects than male proxies do. Those who provide more care for the person have higher ratings of disability of the subject than those who do not provide care. The gender issue may be tied to the fact that women are also providing care.

Should there be a rule that if one is going to be doing a study in which proxies will be considered, the researchers should have a subsample in which they interview both proxies and subjects? For participants who cannot self-report, how relevant is that kind of methodology, because they are really different from the participants for whom one can query both subject and proxy pairs? If people cannot report for themselves, how does one know what somebody else would really be reporting for them? Magaziner responded that he does not know empirically whether that would work. Should there be a subsample? Given currently available information, yes, it would be worthwhile. At least it would help with sensitivity analysis or setting some boundaries on what is learned.

For the oldest old, the proxy frequently is more accurate because in many cases people will underplay their disabilities because of fear of being moved from independent living to assistive living or a nursing home. Another factor is that the elderly person may report that he or she is independent if a caregiver is making activities feasible that would not otherwise be feasible.

Magaziner commented that the direction of the discrepancy becomes important in a population survey in which people of all ages are reporting. One wants to know about that 96-year-old person who cannot quite self-report because he or she does not understand the question. When he or she cannot give what a reasonable person would believe is a reasonable reply, one asks the proxy. Often researchers make simple substitutions, but maybe that is not what one wants to do. Researchers do not have an answer, but that is what needs to be addressed if they do not want to lose people in their attempt to obtain information about the whole population, and not just those who can provide an answer for themselves.

Should one be guided by the findings on the characteristics of the proxies associated with discrepancies in selecting people to serve as proxies? The answer is yes, if one can find the perfect proxy. One has to work with what is available in the real world. The choice may be dependent on the question to be asked, and who has the best opportunity to observe the subject? For example, in a nursing home, perhaps the family proxy is not the best person but someone who sees the subject all the time on a daily basis.

In closing, Andrew Houtenville (New Editions Consulting) informed the participants about two research efforts under way—one led by Mathematica Policy Research and the other by New Editions Consulting, both funded by the National Institute on Disability and Rehabilitation Research, and both about proxy response. Mathematica is going to be working on the question of what protocol is best for a given situation, using an experimental design. New Editions Consulting is going to look at administrative data as a third source of information. Work has been done on this by some economists at the National Bureau of Economic Research using Canadian data in the reporting of diabetes in a working-age population. That work did not involve proxies, but it gave the degree of reliability of reporting diabetes as well as an association with the reporting of a work limitation among the working-age population.

3

Potential Methods for Revising Measures to Foster Comparability Across Subgroups

This chapter focuses on potential methods for refining or augmenting current measures of late-life disability used in population surveys to foster comparability across key subgroups. The presentations covered four topics:

1. Performance measures in population surveys
2. Improving patient-reported measurement of disability using item response theory (IRT) and computer-adaptive testing (CAT)
3. The possible use of easily collected biomarkers of chronic diseases to supplement ADLs and IADLs, which may be able to track decline in functionality over the life course and capture change in functionality across thresholds
4. The potential for using time-use data to augment existing measures of ADLs and IADLs

PERFORMANCE MEASURES IN SURVEYS

Jack Guralnik (National Institute on Aging, National Institutes of Health [NIH]) focused on estimating functional status in surveys using performance measures and on identifying points across the spectrum of performance that are associated with self-reported disability in different population groups. He briefly described several studies he has undertaken with colleagues with some new comparisons both across countries and among U.S. surveys.

Guralnik observed that the Nagi theoretical model of the pathway from disease to disability has been very helpful in terms of operationalizing the assessment of steps along the pathway and particularly useful in thinking about where performance measures fit. Certainly, objective measures of performance can be done at several of the steps in the model—impairment, functional limitation, and disability. Impairments objectively measure physiologic functioning. At the final step, disability, one may be observing people in standardized home-type environments. However, performance measures, such as gait speed, chair rises, and pegboard tests, have been used mostly in the domain of functional limitations.

Guralnik offered three performance assessments to illustrate the points made in his presentation: gait speed, the index of mobility-related physical limitations (MOBLI) developed by Lan and Melzer (Lan et al., 2002), and the Short Physical Performance Battery (SPPB).

MOBLI was developed using data from the National Health and Nutrition Examination Survey III (NHANES III), empirically looking at measures that were related to mobility, which include gait speed, chair rises, and a pulmonary function test. The index was then validated in other studies. The components of SPPB include timed standing balance, a timed 4-meter walk, and timed multiple chair rises. This battery was first developed in the Established Populations for Epidemiologic Studies of the Elderly (EPESE) in 1988. SPPB has very good psychometric properties: It predicts mortality, nursing home admission, new disability, and health care expenditures, among other things, and it has good reproducibility. It is sensitive to clinically important change.

One of the issues related to performance testing in general is that the scoring of some tests does not have a way of dealing with people who are unable to perform the task. So it is difficult to know how to handle people who are unable to perform the test. For example, if gait speed is used, what do you do if someone just cannot walk at all? People have approached this issue in different ways, but it is a limitation of performance measures that researchers rarely address. Even in determining why a test was not done, people often fuss with the data—trying to understand if the data are missing because the person really was not able to do the test and so should be scored as a 0 or given a poor score or whether the person simply refused.

Sometimes even refusals can be vague. People refuse because they are afraid to do the test because they know that they are going to be unable to do it. Sometimes the responsibility for a refusal is placed on the examiner, which is a bit unfair, but it is sometimes hard to sort out when the researcher does not know what the data on the performance test mean. One solution to this problem, used in the SPPB, is to create categorical scores that cover the range of functioning and give a 0 score to those unable to do the test.

Gait speed is an important performance measure: It is a simple test, but it is highly predictive, and recently there has been increasing interest generated in this very simple test, with many longitudinal studies showing a clear stepwise gradient of greater risk for mortality with decreasing gait speed. Analysis of data from the InCHIANTI Study, a population-based study in the Chianti region of Italy (Alessandro Ble and Luigi Ferrucci, Longitudinal Studies Section, Clinical Research Branch, National Institutes of Health, Baltimore, MD, unpublished data), shows a graded response for mortality according to quintiles of preferred walking speed. In this analysis, it was demonstrated that the survival curve for persons with cancer actually showed better survival than the curve for persons in the lowest quintile of gate speed at baseline.

In recent work done in the Whitehall Study of British civil servants (Brunner et al., 2009), it was found that gait speed rose steadily across the six employment grades that were used to classify participants, none of whom was poor and all of whom were full-time employees. It was impressive just how sensitive the gait speed was to employment grades, which range from the highest (administrative level) to the lowest (clerical) level. Gait speed is picking up something about the health disparities across this gradient of socioeconomic status in a very impressive way.

The question often asked is whether performance tests can replace self-reports, whether both should be done, or which one should be used in what situations. Most people who work in the field have generally agreed that self-reports and performance tests are really complementary. They are measuring different concepts, different aspects of functioning; there is a fair amount of evidence to support this view. One example is the work in which Guralnik collaborated with David Reuben (Reuben et al., 1990) in which the study population was stratified in two ways—according to self-reports of being independent in mobility and ADLs and according to categories of SPPB—and mortality was studied as an outcome. In the group reporting no disability, there was a clear grading of mortality risk across SPPB scores. The same was true with the group that was dependent in mobility but independent in ADLs. Complementary information is being picked up, and the performance batteries are showing something that is not available from self-reports.

Finally, among the most severely disabled subset of this cohort, those who were dependent in mobility and with one or more ADL disabilities, there were high rates of mortality. Few people in this subset have high SPPB scores, but even across the remainder of the SPPB spectrum, there was not much of a gradient for mortality risk. Therefore, at the very disabled end of the spectrum, performance measures may not be adding much to the estimation of prognosis, but for those with little or no disability, performance measures make a valuable contribution characterizing prognosis.

Performance Measures in Large Surveys

Banks and colleagues (2006) used both gait speed and the SPPB from the English Longitudinal Study of Aging (ELSA) in its 2004 wave. They used cut points for poor functioning of less than or equal to 8 on the SPPB or gait speed of less than or equal to 0.5 meter per second. The cut points, previously shown to be related to high risk of future adverse events, show a clear age effect, with much higher proportions of people in the 80 years and older group being in the high-risk group according to the SPPB and gait speed, as well as a difference between men and women, with women having poorer functioning.

The presentation by Guralnik compared data for SPPB scores of less than or equal to 8 in EPESE, ELSA, and the InCHIANTI study, with substantially poorer performance seen in EPESE. However, because the EPESE was conducted 10 to 15 years earlier than the other studies, the trends in observed performance may mirror the trends toward less self-reported disability. Similar effects were observed for men as well as for women.

Gait speed showed somewhat different results. For gait speed, Guralnik included NHANES III data from 1988 to 1994. The InCHIANTI population showed a substantially smaller proportion of individuals with slow gait speed, for both men and women. Some of this difference may be real; some of it may be that the test was done in a slightly different way. In the InCHIANTI study, the researchers used automatic timers and participants took a step before the timers were tripped, whereas in the other studies, stopwatches were used and the time was measured from a standing start. It is likely, however, that the Italians, who tend to walk much more than Americans, do have less mobility limitation.

In the NHANES III data from 1988 to 1994, an 8-foot walk was measured. The 2001/2002 NHANES did the 20-foot walk but also timed the first 8 feet to make comparisons with the 1988 data. The results showed a large reduction from one time to the next in the proportion of people who have very slow gait speed. However, some of the difference may be explained by the fact that these tests were done somewhat differently, and in a 20-foot walk people may see a longer walk ahead of them and may go faster for the first 8 feet.

Using Performance Measures to Calibrate Self-Reports

Guralnik stated that he found two examples that represent a way of using performance measures of functioning to calibrate responses to self-report items in questionnaires. In the first example, from the World Health Organization (WHO), Iburg and colleagues (2001) used a modeling technique called Hierarchical Ordered Probit Modeling. They used performance tests from the NHANES III and created a vector of performance that they considered similar to a latent variable representing the true underlying level of performance. Then they used the model to look at how different subgroups reported disability at different levels of this background latent variable. This analysis was done both for physician reports and self-reports.

Guralnik and Melzer (Melzer et al., 2004) did similar kinds of analyses using MOBLI, derived from the NHANES, and observed similar results. People who were 60–69 years old did not report disability until they had a poorer level of performance than people who were older. Also, large differences were observed between men and women, with men not reporting disability until reaching lower levels of performance than women. There were also differences in disability cut points by race, with blacks and Hispanics not reporting disability until their background level of functioning was at a poorer level than that of whites with the same level of functioning. For income, people with the highest income did not report disability until their performance was at a poorer level than that of people with lower income people who reported disability. They may be denying their disability, or they are able to compensate successfully for a lower level of functioning. This kind of approach can be very useful. Comparison of U.S. data with those from the Longitudinal Aging Study Amsterdam (Melzer et al., 2004) showed that people in the Netherlands did not report their disabilities until they had more severe levels of background dysfunction. Therefore, the lower levels of self-reported disability in the Netherlands could be explained, at least in part, by this differential reporting as it relates to level of background performance.

In conclusion, Guralnik observed that there are potential applications of performance measures in improving population surveys of disability, particularly in making comparisons across subgroups of a population and for cross-national and cross-cultural comparisons. Trends over time can be directly observed with performance testing, but this will require strict standardization of test administration and quality control procedures to ensure that the tests are administered precisely the same way in every survey. Performance tests can be used to identify high levels of functioning, which cannot be done well with self-reported disability. They can be used to identify nondisabled persons at increased risk of disability, sometimes referred to as preclinical disability. The concept of calibrating self-reports by using a background measure of performance could be quite valuable. It may be that even something as simple as gait speed could be used for this kind of calibration and could be valuable for cross-national studies.

IMPROVING PATIENT-REPORTED MEASURES USING ITEM RESPONSE THEORY AND COMPUTER-ADAPTIVE TESTING[1]

Karon Cook's (University of Washington) presentation covered four topics:

1. A brief introduction to IRT and CAT
2. Description of Patient-Reported Outcomes Measurement Information System (PROMIS), which applies these methodologies
3. Opportunities and barriers to using modern psychometric methods in population surveys and monitoring population trends
4. Envisioning the future and how modern measurement methods might be helpful in advancing disability research

Item Response Theory

IRT models are probability-based models in which both the levels of the trait being measured (e.g., physical function) and the difficulty of the item are located on a common underlying continuum, or "ruler." The probabilities of answering in particular ways to items that ask about the trait being measured are modeled as functions of how much of the trait a person has relative to the difficulty of the items.

Classical test theory and IRT are different in several ways. Commonly used reliability and validity estimates are based on classical test theory, in which scores on measures are usually obtained by manipulating the item scores (e.g., summing to get a total score). In IRT, scores on measures are obtained on the basis of probability functions, not by averaging or totaling item scores.

Another important difference is that, in classical test theory, unlike IRT, variations in difficulty or intensity of items are not accounted for. For example, a shoulder function scale score on an item that asks about throwing a softball overhand 20 yards is weighted the same as an item that asks about using the affected arm to flip a light switch. In IRT, differences in difficulty or intensity are accounted for. In some IRT models, item discrimination is also included.

Yet another difference is that, in classical test theory, the scores are ordinal-level indicators of individual differences. With IRT, especially with the Rasch model, scores at least approximate interval-level measurement. There is a great deal of debate about how well the scores approximate

[1]For a general review of IRT, CAT, and PROMIS, see De Ayala (1993); Ware et al. (2000); Cook et al. (2005); Cella et al. (2007).

equal-interval measurement, but they come closer than classical test theory scores.

Thus, with classical test theory, the focus is on total score or average score or something along those lines; in IRT the focus is on the item response. The IRT approach gives a great deal more flexibility when developing measures because one gets more information about specific items and how they function.

A very important difference is in the area of reliability and precision. Researchers typically say that a measure has a reliability of, for example, 0.89. Intuitively, everyone knows that it is very likely that a measure will measure at different levels of a trait with different levels of precision, but with classical test theory, all one gets is an average. With IRT, one gets an estimate of the precision of a measure for every level of the trait that is being measured, which gives a great deal more information.

Cook explained that IRT is a mathematical model—a probability model. IRT models estimate how likely persons are to respond in particular ways to a particular item, depending on how much people have of the trait being measured, and what the characteristics of the item are (e.g., how difficult). What is not in the model is the total score. With IRT, different people can answer different items yet their scores are estimated on the same mathematical metric.

Information function in IRT is analogous to reliability in classical test theory. The information function has an inverse relationship to the standard error of measurement; that is, when precision is high, standard errors are low, and when precision is low, standard errors are high. Thus, one can identify what ranges of the trait level are measured with more precision and what areas are measured with less precision. IRT information functions can be estimated at both the scale and the item level.

Computer-Adaptive Testing

IRT is the math behind a very important application—CAT. CAT is a process of measuring in which not all available items are administered to any one respondent. Instead, the items chosen for a particular person are based on that person's responses to the previously administered items.

CAT begins with what is called a large "item pool." Then items of the pool are calibrated in advance on the basis of the known characteristics of the items, their difficulty or intensity, or, in some cases, their discrimination parameters. Once the item pool is calibrated, it is called an "item bank."

Cook then described how CAT works. An initial item is presented to a person and that person responds. Then a gross estimate is made of that respondent's level of the trait being measured. On the basis of that trait-level estimate, the next item chosen from the item bank is the one that gives maximum additional information. Each item has its own information function, and the computer algorithm identifies the item that will give the greatest additional precision for the person's particular trait level. The test administration stops when a predefined "stopping rule" is reached, such as stopping when a person gets to a certain level of precision or stopping after asking a specific number of items.

CAT is not the only application of IRT. It is important to know that with a calibrated bank of items, static instruments can be developed, that is, instruments in which everyone answers the same items. Also, an exciting application is the ability to construct different short forms that target specific clinical populations or specific measurement contexts. For example, one might want a very short measure if respondent burden is an issue. One might want a longer one if precision is of more interest. One might want to target lower levels of the outcome if that is the population of interest, or one might choose items from the bank that seem particularly relevant for a given clinical population.

The important thing to know about item banks is that whether one administers the items with CAT, a long, static instrument, or one of several short forms, the scores are reported on the same mathematical metric. They are not based on a total score but rather on a probability function that takes into account a person's level of trait and responses to items and the characteristics of the items.

Patient-Reported Outcomes Measurement Information System

Patient-Reported Outcomes Mesurement Information System (PROMIS) is an example of an application that uses IRT and CAT; it is funded by NIH. The goal was to develop item banks that measure patient-reported outcomes (PROs) across many different chronic conditions. The focus was on PROs, such as pain, fatigue, physical function, social function, depression, and sleep. Part of the mandate also was to create a computer-adaptive system for administering PRO-based tests to measure such outcomes. The item banks that were developed have the flexibility to create multiple short forms to measure the same traits on the same metric.

Cook explained that one of the things that is most helpful about PROMIS is that scores on all measures have been calibrated to the general U.S. population. For example, for fatigue scores, the mean for the U.S. population is based on a weighted sample that is based on the U.S. census: The mean is 50 and the standardization is 10. Suppose one gives the PROMIS fatigue measure to a particular sample and the average score is 60. This score has inherent meaning: In comparison with the general population, the study population is one standard deviation above the mean. This is much better than using traditional measures, in which scores usually have meaning only to people who have long experience using them.

Opportunities and Barriers

Cook asked: Are these methods—IRT and CAT—appropriate for surveying populations and for monitoring trends? She said yes; in fact, these particular methods offer some distinct advantages, such as the item-level approach, that are very helpful in developing better measures. With trait-specific standard errors, one knows how well one is measuring portions of the population. CAT offers measurement efficiency, and it can be administered in a lot of different ways—by Internet, by telephone, or in person.

IRT also allows linking new instruments to legacy instruments through concordance tables. If two measures are measuring the same trait, it is possible to link them and do a crosswalk between them so that the results of two studies can be compared, even if they used different measures for the same trait.

Cook noted, however, that there are downsides to using IRT and CAT. One is that calibration to an IRT model requires specialized and not particularly user-friendly software and specialized expertise. Also, to use CAT, for example, the respondent or an interviewer has to interface with a computer. If it is an interviewer, then mode effects are introduced that might be problematic. Also, unique qualities of IRT-based measurement require meeting assumptions of the model, and these are not always easy to meet.

Challenges in disability measurement with these particular models are substantial, and so are the advantages. Some of the disadvantages are not limited to the newer methods, however. Because IRT and classical test theory assume unidimensionality, both are probably better suited to measurement of functional limitations than of disability. Disability often gets defined as a multidimensional, interactional, and social construction. Defined as such, it does not lend itself to either IRT or classical test theory methods. Functional limitations typically are defined in much narrower terms and are better suited to measurement models.

Envisioning the Future

In summing up, Cook noted that the psychometric methods that have been developed in the past few years have improved exponentially and have increased researchers' ability to develop good measures in terms of psychometric properties. However, the ability to assign any kind of meaning to those scores is lagging behind. That is an area in need of some efforts. Norm referencing is one possibility, but a great deal needs to be done in addition. Levels and changes in levels of outcomes associated with maintenance of capacity and onset of disability need to be identified. The newer psychometric methods she has been discussing have some unique properties that make them suitable for this endeavor, but they will require longitudinal studies and close monitoring of functions, outcomes, and identification of measurable "marker" clinical events that are associated with changes in PROs.

USE OF EASILY COLLECTED BIOMARKERS OF CHRONIC DISEASES

David Weir's (University of Michigan) presentation focused on the possible use of easily collected biomarkers of chronic diseases, to supplement ADL and IADL measures, which may track decline in functionality over the life course and capture changes in functionality across thresholds. He based his remarks on the results from the 2006 major redesign of the Health and Retirement Study (HRS).

HRS is a longitudinal survey of 22,000 Americans over age 50, who are interviewed every 2 years. Prior to the 2006 redesign, HRS was primarily a telephone survey conducted every 2 years, which included fairly useful self-reports on, among other items, functional limitations, chronic conditions, and care received. Beginning in 2006, the sample was randomly split into two halves: in one, participants are interviewed in person every 4 years beginning in 2006; in the other, they are interviewed in person every 4 years beginning in 2008. The in-person interviews include anthropometric measures, performance measures, dried blood spots, DNA samples, and some other measures not included in the telephone interviews.

Weir explained that the scientific focus of HRS is on two main areas. One area is biomarkers in the narrow sense of biological samples, which are focused essentially on measures of cardiovascular risk. Those biomarkers are relatively easy and straightforward to measure and are of great importance and high prevalence in the population, such as blood pressure, cholesterol, hemoglobin A1c measure of blood glucose, C-reactive protein, waist size, height, and weight. They are also closely related to obesity and metabolic syndrome, which are looming public health concerns. The second area of focus is on physical performance measures, which are targeted more at the older population as measures of frailty.

These two areas can be brought together by taking into account the relationship between chronic disease and disability. Most disability is a product of chronic disease. The chronic diseases that directly produce disability, such as stroke, heart disease, and cognitive impairment, are themselves often produced by antecedent other conditions (e.g., hypertension, diabetes) that often have few symptoms. There is a need to model these processes, as suggested in a previous presentation, beginning long before people have difficulties with ADLs or IADLs. To understand the total process by which people become disabled, it is necessary to look at the whole evolution of chronic disease.

Disability starts long before a person experiences limitation in ADLs. Some measures that are sensitive at those earlier stages are needed. In HRS, there are 12 items, which include such Nagi items as walking several blocks, climbing stairs, and pushing a heavy object. These items are quite useful at documenting the earlier stages of disability. The HRS data indicate that ADL and IADL limitations really only begin around age 75. There are some people with limitations at earlier ages, but these cases mostly do not reflect changes by age. Rather, ADL and IADL limitations are really a feature of the very old. The percentage of people who receive more than 1 hour of care per day also is very low prior to about age 75; after age 75, that percentage increases very rapidly. However, when people who report having no ADL or IADL limitations and therefore are not reporting any hours of care are asked how many of the Nagi limitation items they have any difficulty with, the percentage also rises with age in a very linear way. If people who reported no ADL or IADL difficulties in 2004 are arrayed by the number of Nagi limitation items they had in 2004, and then are arrayed by having an ADL or IADL difficulty by 2006, a very graded relationship is seen. Just counting the number of these difficulties provides some insight into the people who are at risk for developing further disability.

Chronic Disease and Disability

Weir used a combined measure that is a sum of the Nagi items plus ADL limitations plus IADL limitations discussing chronic disease and disability.

As stated above, chronic disease underlies most disability, even at younger ages. People under 62 years of age with disability that prevents them from working are eligible for Social Security Disability Insurance (SSDI). The distribution of SSDI recipients by the cause of disability on which the disability award is based shows that injuries are less than 5 percent and infectious disease about 2 percent. The percentage of recipients reporting diabetes is about the same as injuries overall (4.6 percent). Cardiovascular disease is twice that big (10.4 percent), and arthritis is 2.5 times higher (23.6 percent). All psychiatric conditions, of which depression is the largest, are the single largest cause of reported disability for SSDI respondents (27.9 percent). Even at younger ages, at which most people might think of the disabled as being physically injured, most of it results in some way from chronic conditions. Psychiatric conditions are actually quite important even at younger ages.

In the HRS population, the number of physical limitations rises linearly with the number of chronic disease diagnoses regardless of age, although at 75 years and older the number of limitations reported increases slightly, even at the same number of chronic conditions.

Obesity and Disability

Weir stated that the relationship between obesity and disability is complex. One needs to consider multiple measurement perspectives on obesity. There may be direct effects on mobility: For example, it is harder to move around a lot of weight than a little weight. A somewhat less direct effect is cumulative stress on joints from managing the excess weight. Even less direct effects are through the risk of cardiovascular disease, which may take a long time to manifest itself. There are also inverse effects, such as sarcopenia (the age-related loss of muscle mass, strength, and function) and weight loss.

In HRS, a number of indicators were developed to measure quintiles, with 5 percentile cuts through the variable and then analysis of the mean number of limitations at that level of variablity. Disability and weight as measured by body mass index (BMI) are nonlinear. The lowest 5 percentile is very disabled; this is the frail group. There is relatively little variation for most of the U.S. population, indicating that a range of moderate obesity has relatively low correlation with disability. The number of limitations starts to increase at about the top 25 percent of BMI and especially at the very top—BMIs of 35 and higher.

With regard to the measurement of height and weight, Weir said they added little to the information from self-reports. In fact, they have almost no value analytically. However, it is important to have those data, as they add to age and gender as a predictor of disability.

Waist circumference has a more monotonic relationship with disability. For waist sizes larger than about 40–41 inches, there is a substantial increase in disability. This measure adds considerably to BMI alone. Even with both in the model, it adds significantly: That is, waist size is independently a highly significant predictor of disability. What is it measuring that BMI is not? Weir suggested that it may be central adiposity, which is an independent risk factor for cardiovascular disease. Also, BMI does not distinguish lean body mass from fat. Lean body mass is almost certainly protective in many ways, particularly against mobility difficulties.

Physical performance measures are highly correlated with self-reported limitations. Grip strength, expiratory volume, and timed walk are independently associated with limitations. HRS for a long time has measured cognition, and it is independently predictive of disability, particularly if IADLs are included. A word recall measure has been included in HRS since 1993 and is useful. However, the strongest predictor among the cognitive measures is the eight-item count of depressive symptoms. Depression and other psychiatric symptoms are a major cause of disability. However, they are correlated negatively in self-reports. Separating what may be some kind of affect in a person's reporting style from what is the real effect of depression on disabilities is difficult.

Biological Biomarkers

HRS had high levels of cooperation for collecting biological samples: 80 percent of respondents agreed to do them. The distributions showed a good match to the distributions from the NHANES, except for two measures. One was total cholesterol, which is a difficult assay to do in dry blood spots, and the other was diastolic blood pressure, which is almost certainly due to the fact that machines and humans find that point differently. The biomarkers have good internal validity; prospective validity is to be determined over time.

Disability is only slightly related to current levels of blood pressure, and only at the high end. In contrast, Weir said that quite a strong relationship exists between disability and hemoglobin A1c measures of blood glucose. Disability is correlated with obesity. It has some independent value even after the waist circumference and other obesity measures are taken into consideration.

"Good cholesterol" (high density lipoprotein, HDL) is associated with lower disability. However, disability also is negatively correlated with higher values of total cholesterol, which is quite puzzling. Consequently, a common measure of risk, the ratio of total cholesterol to HDL cholesterol, is not related to disability, at least on its own.

The true value of these blood assays is yet to be determined, in part because a few more assays are still to be done from 2006 data. One is C-reactive protein, which is a marker of inflammation and which may be related to disability through both arthritis and cardiovascular disease. Another is cystatin C, a measure of kidney function. And because blood assays are predictors of cardiovascular disease progression, they are expected to predict future cardiovascular events, which then are precipitators of disability.

DEVELOPING MEASURES OF TIME USE TO STUDY DISABILITY

Vicki Freedman (University of Medicine and Dentistry of New Jersey) described time-use measures and how they may be used to study disability. She also shared some of the lessons from the development phase of a time-use pilot study that she and her colleagues at the University of Michigan are developing for the Panel Study of Income Dynamics (PSID) with funding from the National Institute on Aging.

Her presentation included an overview of three issues:

1. How do time-use data fit in with existing measures of disability (ADLs and IADLs) and some of the conceptual frameworks discussed in this workshop?
2. What are the various approaches for measuring time use to study disability in population-based surveys?
3. What lessons have been learned from the development phase of the PSID's pilot project, Disability and Use of Time (DUST)?

It is not immediately evident how time use fits in with the existing measures of disability. In the Institute of Medicine's (1991) model of the disablement process, conditions and impairments may or may not lead to functional limitations, which in turn, depending on the environment, may or may not lead to disability. It also is not clear how time use fits in with the parallel language offered by the more recent International Classification of Functioning, Disability and Health (ICF) model from the WHO (World Health Organization, 2001). In the ICF model, health conditions may or may not lead to impairments in body functions and structures that may in turn lead to activity limitations and participation restrictions. However, unlike the IOM model, ICF also offers a set of positive analogs for describing functioning. In positive language, the ICF links body functions and structures to activities and participation in daily life. Time-use measures convey the latter concepts: what people do (activities) and the extent to which they engage in social, productive, and other aspects of daily life (participation).

Domains of Time Use

There is no consensus in the literature about how best to classify activities, but if one looks across literatures related to aging, time use, and participation, several key "domains" emerge:

- Basic self-care activities (includes ADLs and other activities that people do to care for themselves, such as management of chronic conditions)
- Household maintenance activities (includes IADLs and other household-related activities that are essential for daily life)
- Regenerative activities (includes hobbies, arts, music, gardening, puzzles, taking classes, etc.)
- Physical activities (includes exercise, walking for pleasure, participating in team sports, etc.)
- Social participation (includes socializing with friends and family, attending group functions)

- Productive participation (includes work, volunteering, providing child and adult care, etc.)
- Political or civic participation (includes involvement in home associations or board meetings, political participation involving collective decision making, etc.)

Not all activities fall uniquely into one of these categories or into just these categories, but these are some of the most common domains of time use for older adults (see Waidmann and Freedman, 2007, for frequency of participation in these types of activities).

Approaches to Measuring Time Use

Freedman explained that there are three main approaches for measuring time use to study disability in population-based surveys. The first is a *24-hour diary*. In such an approach, people are asked a series of questions about everything they did yesterday. The American Time-Use Study, conducted by the Bureau of Labor Statistics, for example, asks respondents what they were doing starting at 4:00 a.m. the previous day, for how long they did it, where they were, and who else was present. The respondents are then asked what they did next, and so on, until a 24-hour diary is completed.

A second approach asks questions about how much time was spent on various types of activities over a longer period of time. These questions are referred to as *stylized time-use questions*. For example, a stylized question might ask: "During the past week how much time did you spend _____?" The reference periods are typically a week or a month or sometimes longer if the activity is rare. This approach can capture activities that are not done frequently.

A third approach, *experiential sampling*, involves contacting study participants at random times of day (with either phones, beepers, or personal digital assistants [PDAs]). The participant is then asked questions about what she or he has been doing in a brief window (e.g., 15 minutes) just before the contact. Depending on the technology, the respondent either answers the question by phone or perhaps types the answers into a PDA. The participants may be asked not only what they have been doing, but also who they were with, where they were, and how they felt.

These three different approaches to collecting time-use data have different strengths and weaknesses. The relative cognitive demands on the respondents vary with each approach. Questions about activities obtained through experiential sampling methods, for example, likely impose the least demand on cognitive skills because of the focus on an immediate time frame. At the other extreme, stylized questions often impose relatively greater cognitive demands, because respondents need to review a longer time period and may need to add or multiply or come up with averages to obtain the number of hours in a week, month, or year for an activity. Somewhere in between is the approach of a 24-hour diary. Another key feature that varies across these approaches is the ability to add descriptors about each activity, such as who the respondent was with, where he or she was, and how the person felt. These questions can be added easily to both the 24-hour diary and the experiential sampling methods; they cannot be easily incorporated into the stylized question approach.

Reliability and validity issues also differ somewhat for each of these approaches. For the 24-hour diaries, for example, weekday and weekend patterns of time use differ. There is also considerable within-person variation across weekdays. Consequently, unless multiple diaries are collected for every person, the 24-hour diary approach is better suited for analyzing population patterns and trends than for analyzing within-person trajectories as people age.

With stylized questions, there are tradeoffs between reference period and accuracy, with the length of the window inversely related to measurement error. One option to minimize measurement error is to use as recent a reference period as possible (e.g., a week) and to focus only on commonly occurring activities.

The experiential sampling approach presents a potentially interesting analytic issue for studying the implications of functioning for time use. Contacting respondents at a specific time and asking them what they are doing yields oversampling of longer lasting activities. If type and length of the activity vary by a respondent's level of functioning, it is possible that bias can be introduced into comparisons of activity duration by functional status. Freedman noted that there are well-established techniques for analyzing length-biased samples, but it is not clear that they have been applied to study disability and time use.

Disability and Use of Time Development Phase

The purpose of the DUST project[2] is twofold: to study the relationship among functioning, time use, and well-being among older couples and to lay the groundwork for potentially collecting time diaries with all adults in the PSID. Approximately 1,600 time diaries will be collected by telephone from 400 married couples aged 50 and older in 2009. Spouses will be interviewed about the same days. Couples will be interviewed about

[2]The DUST project is being led by Vicki Freedman, Frank Stafford, Norbert Schwarz, and Fred Conrad with funding from the National Institute on Aging (P01-AG029409 to Robert Schoeni).

both a randomly selected weekday and a weekend day so that four diaries per couple will be completed in all. For each spouse, the first interview will also include supplemental questions to assess stylized time-use questions, detailed measures of functioning, and global and detailed measures of well-being.

The DUST team has spent almost 2 years developing the instrument. The development phase included a series of focus groups, cognitive testing of the instrument, an assessment of the reliability of diary pre-codes,[3] and a pretest with 27 couples. In terms of questionnaire design, the team began with the American Time-Use Study questions, which ask respondents what they were doing for how long, who was in the room with them, and where they were. DUST investigated several expansions, which included

- The distinction among who actively participated in the activity with the respondent, who was there but not actively participating, and for whom the activity was carried out. This involved the testing of nine pre-codes that route respondents to different ("tailored") follow-up questions depending on the type of activity reported.
- Introduction of a tailored follow-up to determine whether the respondent received help with each reported activity or did it on his or her own.
- The addition of a single-affect measure for each activity in the diary that correlates well with established measures of well-being from diaries such as the Day Reconstruction Method (Kahneman et al., 2004) and the Princeton Affect and Time Study (Krueger and Stone, 2008).

Freedman reported that several useful lessons about time-use diary measurement have emerged from the development phase of DUST.

The first lesson is that activity descriptors about help may not be consistently interpreted. Focus group activities suggested that adding a follow-up question about receipt of help with each activity might yield inconsistent responses. Interpretation of what constitutes "help" varied and was related to the couple's division of labor and the spouse's ability to carry out such activities. This lesson was learned early in the development phase, and therefore this line of questioning was dropped prior to cognitive testing.

The second lesson is that pre-codes to tailor descriptors can be reliably incorporated into the time diary. To tailor descriptors to different types of activities reported in the 24-hour diary, the team piloted nine pre-codes to

[3]The term "pre-code" is used to distinguish from the type of coding that more typically occurs after the diaries are collected (i.e., post-processing or post-coding). Both types of coding will be done in this project.

be coded during the interview. For example, for household chores, helping, and care-related activities, a follow-up question, "Who did you do that for?" was asked, along with questions about who did that with you, and who else was there with you. The interrater reliability of selecting one of nine pre-codes in two rounds of testing with four interviewers was very high (kappa > 0.9). Furthermore, in pretest interviews, which yielded over 1,500 activities, interviewer pre-codes agreed with the coding of the principal investigators more than 90 percent of the time.

The third lesson is that a succinct measure of well-being can be incorporated into the diary as a valid activity descriptor. The team developed and tested an activity descriptor to tap affect for all activities reported during the previous day. In focus groups, respondents were asked an open-ended question about how they felt for each activity reported during the previous morning. Participants were then asked to classify these emotions as mostly unpleasant, mostly pleasant, or neither. Participants were able to classify their emotions in ways that made sense, but they needed direction in cases in which they experienced both positive and negative emotions. From this experience, the team developed the following question: "How did you feel while you were ___? If you had more than one feeling, please tell me about the strongest one. Would you say mostly unpleasant, mostly pleasant, or neither?" Based on the pretest data, the correlation between responses to this item and to more detailed questions about activities that occurred during three randomly selected times of day, which were modeled after the Day Reconstruction Method, was relatively strong (> .7; N = 155).

The fourth lesson is that less cognitively demanding stylized questions can be successfully administered to couples. Rather than asking how much time respondents spent in the last week or month doing specific kinds of activities, DUST included in its cognitive testing and pretest questions of the form: "On how many of the last 7 days did you ____?" Respondents were provided with the following categorical answers to choose from: none, 1–2, 3–4, 5 or more. Every one of the participants in the cognitive testing was able to answer these questions. When asked how they arrived at their answer, some participants reported knowing their schedules and others reported reviewing and counting each day in the previous week that they performed the activity. No problems were identified with these items in subsequent pretesting.

The DUST team anticipates making the data available for public use by the end of 2010 on the PSID website. The pilot will offer not only a larger sample size, but also multiple days for each person and same-day diaries for couples so that investigators can explore a number of crucial questions related to older couples' functioning, time use, and well-being.

DISCUSSION

Participants asked several questions for clarification or elaboration, mostly focused on four topics: PROMIS, CAT, time-use measures to study disability, and analysis of late-life disability.

Patient-Reported Outcomes Measurement Information System

Several questions were asked about PROMIS. The ability of PROMIS measures to distinguish people who score very low on a trait, such as physical function, was discussed. Cook explained that PROMIS item banks are developed so that there are items that target both high and low levels of the trait. Large item banks are potentially better at discriminating among frail elders, for example, because there are many items that match their levels of function. Although PROMIS scores are normed (i.e., the average scores in the general population are known), PROMIS measures still do a good job of measuring people with extreme levels of a trait (e.g., very low physical function).

Norms for different age groups and different clinical and social groups can also be calculated. For example, PROMIS has calculated the average scores for persons in different age categories, for gender, for different clinical conditions, and for persons with none, one, two, or more chronic or disabling conditions.

Computer-Adaptive Testing

A clarification was made about the difference between CAT and screening questions. Researchers studying trends in disability worry that a screening question might prevent getting information about the prevalence of something asked about in a follow-up question. Fortunately, the items that are presented with a CAT are not screened; they do not keep someone from being asked about some other condition; they only help decide which questions will give the most information about someone's level of, for example, mobility.

A potential problem with CAT was mentioned—the whole idea of "framing"—the phenomenon in which one question on a measure might cause a person to think about the rest of the questions in a particular way (framing). Researchers realize that the question that is asked beforehand impacts how one answers a particular question. This is a serious issue to consider with CAT. Short forms do not have this issue as much, or rather, they have this issue, but it is the same for everyone taking the test.

Time-Use Measures to Study Disability

Participants discussed the point made in the session that a more disabled person might take longer to do something than a less disabled person. In experiential sampling, that person is more likely to be picked up doing the specific activity than others. Is that a length-biased sample of activities or a measure? The issue is what is being measured—the proportion of people doing activities and the length of time on activities are two different questions.

Another use of the time-use method is to track changes in patterns of activity or participation over time, which may reflect changes in health as well as disability status. But at a point in time, how does one determine what is normative and what may be reflective of poor health or disability? Still another issue raised concerns how to interpret the responses if one asks people on how many of the past 7 days they have done some activity that is considered elective. They may choose to do it or not do it. How does one know whether the decision to do it or not to do it is related to their functioning or that they are just not interested in the activity, such as socializing or going to meetings? There are a couple of ways to answer that question: asking people whether they do an activity as much as they like to, as well as linking it to health-related reasons; or asking people what it is that they value and then tracking their participation in those activities. Questions can be individualized to what people say is important to them.

Analysis of Late-Life Disabilities

A participant commented that late-life disabilities are a manifestation of the life-long accumulation of activities. The data now available in the United States do not allow a life-course study of how early exposures to negative factors in personal traits, and also the environment, result in any late-life disabilities. The earliest data available on late life are maybe from HRS.

Guralnik observed that in contrast to the situation in the United States, the British have birth cohorts, the oldest of which is now over 60 years old. They also have cohorts that started a little bit later in life that are now aged. The evidence is clear that early life factors play a very large role in mid-life and late-life functioning. Participants agreed that such cohorts are invaluable for studying the life-long development of disabilities. In some cases, existing cohorts could be used if there are mid-life or earlier data and one can recontact people when they are older. In that vein, it was noted that one of the rationales for adding the study of disability to the PSID is that the panel study is over 40 years old now and does have some predictive measures, mostly economic ones, of distress in early life.

4

Improving the Validity of Cross-Population Comparisons

This chapter discusses methods for improving the validity of cross-population comparisons, within and across countries, for measures of disability obtained in population surveys. The presentations covered three issues:

1. Developing additional measures of limitations in cognitive functioning and disability that could be used in population surveys
2. Using vignettes for validating judgmental reports in population surveys
3. Approaches to cognitive and field testing of disability measures for cross-cultural and cross-national comparability

ADDITIONAL MEASURES OF LIMITATIONS IN COGNITIVE FUNCTIONING AND DISABILITY

Craig Velozo's (University of Florida and the Veterans Affairs Medical Center in Gainesville) presentation addressed the relationship of limitations in cognitive functioning and disability, additional measures of cognition, and item response theory (IRT) and computer-adaptive testing (CAT). Velozo explained that the issue of cognition is very relevant to disability among the elderly population. A quick review of the literature shows a positive relationship between cognitive function and ADL and IADL status (Barberger-Gateau et al., 1999; Steen et al., 2001); decreases in ADL and IADL performance associated with cognitive decline and mild cognitive impairment and disability (Di Carlo et al., 2000; Kumamoto et al., 2000;

Purser et al., 2005; Raji et al., 2005; Ishizaki et al., 2006) and cognitive decline and ADL limitations associated with increased mortality (Wu et al., 2004; Schupf et al., 2005).

Current Measurement

Velozo pointed out that the following cognitive instruments are typically used in national population surveys:

- The Mini Mental State Exam (MMSE): 11 questions covering 5 areas—(1) orientation, (2) registration, (3) attention and calculation, (4) recall, and (5) language
- The Medical Expenditures Panel Survey instrument: questions addressing memory loss, confusion, problems making decisions, and supervision for safety

Cognitive instruments generally used in rehabilitation include

- the Functional Independence Measure (FIM), used for inpatient rehabilitation, has five questions that address memory, comprehension, expression, social interaction, and problem solving;
- the Minimum Data Set, used in skilled nursing facilities, has approximately 11 questions that address long-term memory, short-term memory, daily cognition, awareness, and speech and understanding; and
- the Outcome Assessment and Information Set, used in home health, contains a subset of questions that are somewhat cognitive and somewhat leaning toward function, such as managing oral medications, using the telephone, cognitive function, and speech clarity.

Velozo said that these instruments have some limitations, both in content and in measurement. Relative to content limitation, MMSE does not address the effects of cognition in a person's daily life. MMSE also does not generate separate cognitive domain measures that are more typical in the neuropsychological literature, such as attention, memory, and executive function. Relative to measurement limitations of cognitive assessments, although FIM is widely used and has a relatively extensive literature on its psychometrics, these psychometric studies focus on the "motoric" or ADL component of FIM, not the cognitive component.

Recent developments in the area of "applied" or "functional" cognition offer one of the potential solutions for content limitations. Coster and colleagues (2004) have defined applied or functional cognition as discrete functional activities whose performance depends most critically on the application of cognitive skills with limited movement requirements: for example, daily activities that require cognition, such as finding keys; conversing with more than one person; and resolving a simple problem, such as scheduling a doctor's appointment. They developed a measure of applied cognition that includes 59 items, which are based on the International Classification of Functioning, Disability and Health (ICF). These investigators tested the items on 477 patients who were receiving rehabilitation services. They applied Rasch measurement (an IRT methodology) and used principal components analysis (PCA) to investigate the unidimensionality of this set of items. Of the 59 items, 46 fit the Rasch model; 25 percent of the sample was at the ceiling; and the PCA suggested that the instrument was unidimensional.

In contrast to the available traditional cognitive measures, Coster and colleagues (2004) used an IRT approach that involves the use of relatively large item banks to measure individuals. Associated with the IRT is the calibration of items according to their "difficulties" (see Chapter 3). Velozo stressed that the calibration is an important aspect of the IRT approach that offers some benefits in terms of understanding the measures. (As discussed in Chapter 3, IRT is the statistical foundation for CAT, which is a method to administer subsets of items that are individualized for the respondent.)

A New Applied Measure

Velozo described his work with colleagues in which they used IRT and CAT in the development of an applied measure of cognition for stroke patients. The purpose of the study is to develop a measure of cognition that reflects the impact of cognitive challenges in everyday life; to design measures for separate domains of cognition (e.g., attention, memory, executive function); and to maximize measurement efficiency and precision using IRT approaches and CAT. This work involved two studies: (1) developing a Computer Adaptive Measure of Functional Cognition (CAMFC) for Traumatic Brain Injury and (2) developing a similar measure for stroke.

Velozo gave an overview of the stroke study. Although it did not include typical aging individuals, within the stroke population there are individuals who have no or fairly mild cognitive problems and so may be reflective of what might be seen with an aging population.

The four steps in developing a measure of functional cognition for stroke patients were as follows

1. Develop domains of functional cognition (Donovan et al., 2008), with input from an advisory panel on initially proposed domains.
2. Develop an item pool of cognitive items, using focus groups that included health care professionals, patients, and caregivers for the initially proposed sets of items.

3. Field test the item bank, using confirmatory factor analysis (CFA), Rasch psychometrics, and correlations with neuropsychological and functional assessments.
4. Develop a CAT version of the measure.

The 10 final domains of functional cognition included language, reading and writing, numeric calculation, limb praxis (which is very specific to the area of stroke), social use of language, visuospatial functioning, emotional function, attention, executive function, and memory. Operational definitions were developed for each of these domains. Each domain had subsets of items; the number of items per domain ranged from 9 to 41. An item pool was developed for each domain, which resulted in 244 functional cognitive items across the 10 domains.

A total of 128 individuals were tested: 49 were acute stroke patients and 79 were chronic stroke patients. Psychometric analysis was performed on 252 ratings: 128 were self-ratings and 124 were proxy ratings from caregivers. Concurrent validity (CAMFC-Stroke domains against neuropsychological-functional test) was investigated on a random selection of 63 participants. CFA supported treating the 10 domains as a "single measure." Except for the limb praxis domain, the majority of correlations across domains were in the moderate range. CFA of each domain provided mixed results in supporting the unidimensionality and hypothesized multiple-factor structure of the domains: 5 of the 10 domains showed support for both a unidimensional factor structure and a multiple factor structure (based on neuropsychological subdomains); 1 of 10 domains showed support for only a hypothesized multiple-factor structure; and 4 of 10 domains failed to support either a hypothesized undimensional or a multidimensional structure.

A single measure across all domains (as rated by patients) and the domain measures (with the exception of limb praxis) showed a high percentage of items fitting the Rasch measurement model. Both the single measure and the domain measures (except limb praxis) showed good internal consistency and construct validity.

With regard to measurement sensitivity, the single measure showed excellent sensitivity in separating the sample into different "ability" levels. Except for limb praxis and numeric calculations (patient-reported), domain measures showed good sensitivity in differentiating the sample. The single measure showed no floor or ceiling effects. The domain measures showed no floor effects, but 5 of the 10 domains showed ceiling effects.

Results of rater comparisons showed that patient self-reports correlated with the caregiver proxy reports in the fair to moderate range for all domains except limb praxis. Patients and caregivers rated items in a similar way, as indicated by low levels of differential item functioning (DIF).

With regard to concurrent validity, domain measures showed fair to moderate correlations with analogous neuropsychological-functional tests. Caregiver proxy reports had a tendency to show stronger correlations with analogous neuropsychological-functional measures than patient self-reports.

In conclusion, Velozo said that CAMFC-Stroke can exist as a single measure or as a battery of nine domain measures (excluding limb praxis). The advantage of the single measure is excellent measurement sensitivity, and the advantage of the domain measures is the ability to monitor domain-specific outcomes. In their study, both patients and caregivers provided acceptable CAMFC-Stroke measures.

The item difficulty hierarchy within each domain of the CAMFC-Stroke offers considerable information. It provides support for the hypothesized domain development structure and provides a basis for interpreting the measures that are generated. Unique to this kind of measure is the capability of interpreting the generated measures in terms of what the patient can and cannot do within the content of the measure. For example, for a person who receives a measure of 0 logit, items such as "copies information correctly" and "pays attention to an hour-long TV program" should match his or her ability level; items at –0.75 logit (such as "correctly answers yes/no questions" and "greets someone who enters the room") should be easy for the individual; and items at 0.75 logit (such as "has a conversation in a noisy environment" and "reads 30 minutes without taking a break") should be difficult for the individual.

In summary, newly developed measures such as the CAMFC-Stroke extend the capability of measuring cognition on several fronts. First, IRT approaches maximize precision by generating measures from groups of items (i.e., item banks). Second, in combination with CAT approaches, IRT-generated measures reduce respondent burden. Finally, since IRT approaches provide item-difficulty calibrations, measures generated with these instruments can be interpreted in terms of what individuals can and cannot do. While still in their infancy, IRT-CAT approaches to measuring cognition and its impact on everyday life show promise for population-based measurement.

USING VIGNETTES TO IMPROVE CROSS-POPULATION COMPARABILITY OF SELF-RATED DISABILITY MEASURES

Arthur van Soest (Tilburg University and RAND) began his presentation by noting that work-limiting disability is a major problem in many developed countries. It reduces participation and national productivity and increases the social welfare burden. Individuals with work disabilities lose income, and their quality of life is lowered. The problem will increase as the population ages and people retire later in life.

A simple work disability self-assessment—such as "Do you have an impairment or health problem that limits the amount or type of work you can do?"—is often used to measure work disability and compare work disability rates across countries or socioeconomic groups. Large and significant differences between countries that seem to be at similar levels of development are found in self-reported work disability rates.

Van Soest reported on a study comparing workers in the United States and the Netherlands (Kapteyn et al., 2007). In the study sample, 4.9 percent of the people in the United States reported being on disability rolls, compared to 10.7 percent in the Netherlands. There may be several explanations for this difference. First, programs providing disability benefits in the two countries differ in terms of financial incentives, access criteria, and application procedures. Second, people in the United States may be healthier than those in the Netherlands. Third, American employers may accommodate workers with a health problem better than Dutch employers.

In this study, the researchers focused on the second and third explanations. Are Americans really healthier than Dutch workers or are employers in the United States better able to accommodate workers with a handicap than in the Netherlands? The question—"Do you have an impairment or health problem that limits the amount or type of work you can do?"—is a very general measure of work-related disability, and typically is the only question, or some rephrasing of it is used, in general socioeconomic surveys where there is little room for elaborating on each specific topic. The responses to this question show that the prevalence of work-related health problems according to self-reports is much higher in the Netherlands than it is in the United States for all age groups. Are these "real" differences or differences in "reporting style"?

If these are real differences, then one would expect to observe similar differences in the prevalence of chronic conditions that may lead to work-related health problems. Some examples are diabetes, arthritis, hypertension, heart problems, stroke, and emotional problems. However, a comparison for the age group 55 to 64 shows that people in the Netherlands actually suffer less from chronic health conditions than people in the United States. This finding suggests that the two countries may differ less in measured work disability than is reported by individuals.

Van Soest then reported on research using anchoring vignettes in his and his colleagues' survey in the Netherlands and the United States. The methodology for this work was based on the earlier work by a group at Harvard in cooperation with the World Health Organization (WHO) (King et al., 2004). They found that reporting differences explain more than half of the observed differences in self-reported work disability, leaving less than half as differences in underlying real health or employer accommodation of workers with a handicap. So looking only at self-reports one will draw misleading conclusions. Correcting for differences in the responses is essential in order to compare the actual distributions of health in the two countries. What is needed is to correct for the fact that people in different cultures have different response scales, different norms to say whether they have work-related health problems or not, or different norms for the severity of their work-related health problem.

Self-reports provide a good alternative to the difficult (or impossible) and expensive task of creating a complete and comprehensive objective measure of work disability, but they have the drawback of possible differences in reporting styles. Technically, such differences are called DIF. Vignettes can be used to analyze these differences in response scales. Vignettes are a new experimental tool that can correct self-reports and make them comparable across countries or socioeconomic groups; they work particularly well across countries because those are the comparisons for which differences in reporting styles are largest. A vignette describes the health of a hypothetical person and then asks the respondent to evaluate that person's health on the same scale used for the self-report on health. Since the vignette description is the same in the two countries, the actual health of the person described in the vignette is the same. Therefore, any difference in reported country evaluations must be due to DIF.

Van Soest and his colleagues applied the vignette approach to work-limiting disability to obtain not only international comparisons that are corrected for DIF, but also comparisons of different groups within a given country, such as systematic testing of hypotheses of differences by sex, age, or socioeconomic status. Vignettes were developed in three domains of disability: (1) back pain, (2) mental problems, and (3) cardiovascular disease. Respondents in both countries were presented with these vignettes involving several questions, on a two-point scale and a five-point scale, and asked questions very similar to those asked about themselves. They were asked to evaluate the hypothetical persons presented for each of the vignettes. The response scales were the same as the response scale for the self-reports. The responses were used to estimate several versions of an econometric model generalizing the model introduced in the work of King and colleagues (2004). Van Soest and colleagues found that U.S. respondents were "harder" on the vignette persons than the Dutch respondents: many more U.S. respondents than Dutch respondents said that the vignette person had no or only mild problems with working, whereas many more Dutch respondents thought the person had more serious problems in terms of working.

Using simulations on the basis of the estimates, they found that according to a model not using vignettes, the percentages of people aged 51–64 with a work disability (on a yes/no scale) was 36 percent in the Netherlands and 23 percent in the United States. Correcting for the response-scale differences using vignettes and the benchmark model, the difference was reduced substantially. In the simulation, every respondent was given the U.S. scales. Nothing changed for the U.S. respondents, but the Dutch respondents appeared to have much less work disability when using the U.S. scales than when using their own scales. Accordingly, the model with vignettes and accounting for response-scale differences gave a much smaller difference in work disability between the two countries than a standard model that assumes everyone uses the same scales.

Within the basic structure, van Soest and his colleagues used several models to test the sensitivity of the main result to different model assumptions. Basically, they were all technical changes to the model, and not many changes in the results were observed. They consistently found that vignettes on work-limiting disabilities do help to correct for cross-country differences in scales used in self-reports. Corrections using vignettes reduced the estimated difference in work-limiting disability between the United States and the Netherlands by more than half. This result was robust to specification choices as long as the vignettes on all three domains (pain, cardiovascular disease, and mental health problems) were used.

What explains the remaining difference? That is something the researchers still do not know. Can the differences be explained as employer accommodation? It is possible that employers in the United States are more used to having employees with a disability than employers in the Netherlands, where it is traditional for people with disabilities not to work? The researchers were unable to study the distinction between health and employers' accommodations to it.

Similar studies have been conducted in a number of European countries. In the Survey of Health, Aging, and Retirement in Europe (SHARE), similar questions were asked in eight countries in 2004. The COMPARE[1] subsample of the SHARE 2006–2007 with the same work disability vignettes found that, if everyone uses the same response scale, on the five-point scale the percentages responding that a person has no problem or just a mild work-related health problem are not too different between countries.

Finally, vignettes as a methodological tool can be applied in many other domains. In earlier work, they have been used by the Harvard Group and WHO in the fields of health and health care quality and political efficacy.

[1]COMPARE is part of the family of research projects linked to SHARE. Data collection is parallel to the SHARE data collection in waves 2004 and 2006–2007 and follows the same procedures.

In addition, the American Life Panel,[2] the Dutch CentERpanel,[3] and the COMPARE samples on the age 50 and older populations in 10 European countries have vignettes on satisfaction with income, work, or daily activities, and general well-being.

NEW APPROACHES TO COGNITIVE AND FIELD TESTING OF DISABILITY MEASURES

Julie D. Weeks (National Center for Health Statistics, NCHS) began her presentation by emphasizing that it is nearly impossible to discuss the recent advances by NCHS in question testing and evaluation methods without first considering two international question development projects, because that work has informed and transformed the testing work. There are two characteristics of the question development projects that have significantly influenced the way in which the testing and evaluation methods have developed: the specific desire to *not* rely blindly on existing questions, which may erroneously be considered "gold standards," and the fact that the questions are intended for use in trend analysis and cross-cultural comparative work. Weeks stated that she would describe the question development initiatives and then turn to the impact that these initiatives have had on the way cognitive question testing is conducted now, both at NCHS and at partner sites around the world.

Question Development

At the international level, there is largely an absence of comparable measures that can be used to paint a broad statistical picture of population health and disability. This is not to say that comparisons are not made—there certainly is information on births and deaths and life expectancy used to make general statements about the health of a population—but consistently measured, specific, standardized measures of health and disability status do not exist. Furthermore, standards with regard to the conceptualization, definition, and collection of those measures and the conduct of analyses typically are also lacking.

Under the auspices of the United Nations, national statistical offices, and the Conference of European Statisticians, two groups were formed and charged with developing such measures that would provide basic in-

[2]The American Life Panel is the U.S. analogue of the CentERpanel—it is representative of the U.S. population

[3]The CentERpanel is an Internet survey based on a random sample of the Dutch people aged 25 and older. It is administered by CentERdata, a research institute affiliated with Tilburg University.

formation on population health and disability, for both within-country and international comparisons. Those two groups are the Washington Group on Disability Statistics and the Budapest Initiative. The Washington Group on Disability Statistics operates under the aegis of the U.N. Statistical Commission. Its main purpose is the promotion and coordination of international cooperation in the area of health statistics by focusing on disability measures suitable for censuses and national surveys. The Budapest Initiative, which is formally the Joint United Nations Economic Commission for Europe/WHO/Eurostat Task Force on the Measurement of Health Status, was organized under the Work Programme of the Conference of European Statisticians. Its main purpose is the development of an internationally accepted standard set of questions for assessing general health state in the context of population surveys. An important objective of both efforts is to maximize the cross-individual and cross-population comparability of survey questions and resulting data.

Participants in the Washington Group include representatives from over 60 countries, national statistical offices, international organizations, and nongovernmental organizations, as well as some disabled persons' organizations. Using ICF as a framework, the group's interest is in the measurement of basic actions at the level of the whole person. Disability is defined as the intersection of basic actions and the environment and affects participation in society. Disability is treated as a demographic variable, comparing populations and subgroups by disability status.

The Washington Group first developed a short set of six disability questions that have been tested, adopted, and now are being included in plans for censuses around the world. The group is now engaged in the development of longer sets of questions that include increasingly complex activities and more domains of health.

The major focus of the Budapest Initiative is on the development of measures suitable for population surveys that capture health status or "health state." In this context, health state reflects one's functional ability ("within the skin" as opposed to with the use of aids or other assistance); that is, capacity, rather than performance, in a reasonable environment.

Like the Washington Group, one of the objectives of the Budapest Initiative is to develop a question set that describes individuals' overall health state, by examining functioning in basic levels of activity across a number of health domains. A second objective is to describe trends in health over time within a country, across subgroups of a population, and across countries. In this way, something meaningful about a population's health can be said when examining differences between countries and assessing trends over time.

Weeks noted that the objectives of the two groups are very similar. When their respective work groups mapped out the possible health domains and basic activities that could be measured, the list included six categories:

1. Mobility—walking, climbing stairs, bending, reaching or lifting, using hands
2. Sensory—seeing and hearing
3. Communicating—understanding and speaking
4. Cognitive functions—learning, remembering, making decisions, and concentrating
5. Emotional functioning—interpersonal interaction and psychological well-being
6. Other—affect, pain, fatigue, and self-care

The challenge that each group faced was demonstrating in some easily digestible way exactly where the questions being developed were located in what seemed like an ever-increasing map of health and disability. Both groups are measuring functional ability at the level of basic actions, but across multiple health domains. How does one clearly show what is being considered and what remains to be developed? Moreover, the Washington Group is defining disability as the intersection of the person and the environment, so one has to know something about the environment. Finally, it became increasingly apparent that in a survey setting, in which there is room for a relatively larger number of questions, the question of what additional aspects should be measured within any health domain adds to the complexity. After many months and iterations, a small group of members developed "the matrix." At the simplest level, this matrix outlines in what areas work has occurred and in what areas it needs to continue to develop a full spectrum of questions on health and disability.

The goal is to populate each cell with questions that have been subject to rigorous testing so that countries can use comparable measures and can choose those measures that fit with their survey and budgetary agendas. Particularly noteworthy about this matrix is that it succinctly conveys the use of explicit definitions of disability (or health) and is, in essence, a roadmap for future survey and question development work. However, finding questions that have been cognitively evaluated is nearly impossible. Furthermore, in earlier efforts, it quickly became clear that one simply cannot rely on data collected in separate studies, nor can the findings be compared.

Question Testing and Analysis

The question testing and evaluation phase of work required nearly as much development as did the questions. Ultimately, the goal of both the Washington Group and the Budapest Initiative is to develop internationally comparable data that are suitable for censuses and surveys and that capture *most* disabled people (or the broad spectrum of health states) *in a consistent fashion*. The goal then, for the cognitive test, is to ensure that the questions meet those goals, without relying on a gold standard. Unfortunately, many of the traditional aspects of cognitive testing do not produce consistency or standardization. Furthermore, it is often impossible to assess if differences are "real." Some of the aspects of traditional cognitive testing methods that hinder comparative analysis include small, nonrepresentative samples; nonstandard interviewing protocols, outputs, and reports; underdeveloped literature and practice regarding the rigor of analysis; and lack of standardized criteria for what constitutes a cognitive interview finding.

Also, in the area of disability measurement (as in so many other disciplines), existing questions are often used as gold standards, and new questions are evaluated by examining the relationship between the two. This is a problem for several reasons. First, often the gold standard has not been rigorously tested, so it is not clear what is being compared and which measure might, in fact, be superior. Moreover, the purpose of the gold standard questions and new questions may differ, even slightly, so that making such a comparison may not be entirely appropriate. Finally, the strategy does not address cross-cultural comparability, unless it was addressed in the development of the original question considered the gold standard, which one would never know because questions currently in use rarely come with evidence of such study.

The cross-cultural nature of the Washington Group and Budapest Initiative projects underscores the need to clearly demonstrate that a question works and what is being measured. It is no longer sufficient to know just that the "questions worked"; one needs the question wording, interpretation, and outputs, as well as how respondents interact with and answer the question.

This need required a huge paradigm shift in the cognitive testing lab and ultimately changed the way the testing and the evaluations are being conducted. In effect, the qualitative process is subject to far more of the scientific principles associated with quantitative analysis: a structured cognitive interview, data quality, data analysis (multiple levels of analysis, including an examination of patterns of respondent interpretation and calculation), and transparency and replicability in all processes, but, most importantly, in the transformation of qualitative data into quantitative results. In addition, these methods have to be implemented in a consistent and standardized fashion across all of the participating countries.

Weeks next described the cognitive testing used by the Washington Group and Budapest Initiative. One of the most important steps during the initial phase is stating as specifically as possible the research questions to be tested. Moreover, one does not simply want to know that the "questions worked," but rather:

- How do specific respondents move through the cognitive processes (comprehension, retrieval, judgment, response)?

- How much error is there (false negatives and false positives)?
- Why are there differences and how does one account for the way respondents with different socioeconomic conditions, cultures, and languages interpret, consider, and respond to survey questions?

The challenge of answering these three questions is heightened even further when attempting to design an internationally comparable measure for a concept as complex and dynamic as disability.

Next in the process is putting together a testing work group that meets on a regular, frequent basis. There is an initial, mandatory training meeting for all participating countries (or sites). A purposeful sampling procedure is designed; it is not a convenience sample. Translation is a major activity requiring a great deal of time and care. As much time is spent in translation as is spent in nearly all of the rest of the planning phase of testing. Ultimately, if one is going to administer questions about how "sad, blue, or depressed" a respondent is in one site (the United States, for example) one has to know that these terms mean exactly the same thing in the other test sites. In Italy, for example, there is no concept of the term "blue." Even if one can find the word, does it mean the same thing? Is it going to result in comparable data? Finally, time-intensive work must be done to take notes and to translate those notes into some kind of quantitative format. However, the data that are generated are rich and very quantitatively informative.

When this work is completed, one has not only the typical qualitative notes taken during a cognitive interview, but also a narrative that follows a semi-scripted format, with as much detail as possible. In turn, those narratives are entered into QNotes (software designed at NCHS) and form the basis of the data from the cognitive testing that are analyzed in a very quantitative fashion. The results of this process are

- validity tied to rich detail,
- findings that are grounded,
- insight into question interpretation,
- insight into patterns of calculation, and
- knowledge of question performance.

Weeks next described what is different about the analysis stage. In quantitative terms, she and other NCHS staff liken within-interview analysis to frequencies, across-interview analysis to conducting crosstabs, and across-subgroup analysis to controlling for specific variables. The point is that each type of analysis offers some type of understanding. The goal is to perform the type of analysis that answers the question of most interest, but typically all three of them.

The analysis itself should be conceptualized in three distinct layers. The first and simplest level of analysis occurs within the interview, specifically, as the interviewer attempts to understand how one respondent has come to understand, process, and then answer a survey question. The interviewer must act as analyst during the interview, evaluating the information that the respondent describes and following up with additional questions if there are gaps, incongruencies, or disjunctures in the explanation. From this vantage point (i.e., within a single cognitive interview), basic response errors, such as recall trouble or misinterpretation, can be identified—errors that can be linked to question design problems.

The second layer of analysis occurs through a systematic examination of all interviews together. Specifically, interviews are examined to identify patterns in the way respondents interpret and process the question. By making comparisons across all of the interviews, patterns can be identified and then examined for consistency and degree of variation among respondents. Inconsistencies in the way respondents interpret questions may not necessarily mean misinterpretation, but they can illustrate even the subtle interpretation differences that respondents use as they consider the question in relation to their own life circumstances. From this vantage point, it is possible to identify the phenomena that are captured by the particular survey question, illustrating the substantive meaning behind the statistic. Additionally, from this layer of analysis, it is possible to identify patterns of calculation across respondents. This is particularly useful in understanding how qualifying clauses, such as "in the past 2 weeks" or "on average," affect the way respondents form their answer and whether respondents consistently use the clauses in their calculation.

The last level, the heart of the cross-cultural analysis, occurs through an examination of the patterns across subgroups, identifying whether particular groups of respondents interpret or process a question differently. This level of analysis is particularly important because it is the level where potential for bias would occur.

Thus far, this testing protocol has been used by both the Washington Group and the Budapest Initiative in approximately 30 countries. In a subset of these countries, staff has also combined the cognitive testing with field testing. Preparations are now being made for a combined cognitive and field testing effort in the U.N. Economic and Social Commission for Asia and the Pacific region, which will include Cambodia, Fiji, Maldives, Philippines, Sri Lanka, and Vietnam. It has been a remarkable endeavor, one that has produced exciting results in both the development of questions and testing for cross-cultural purposes.

In summary, the Washington Group and Budapest Initiative question development work is located at the most basic levels of activity and participation in core health domains. The goal is to measure ability (or inability) to carry out basic activities and to treat disability as a demographic variable, comparing participation by disability status. In this context, health state reflects one's functional ability "within the skin" (without the use of aids or other assistance), rather than performance, in a reasonable environment. A new methodology for integrating cognitive testing concepts into a standardized, quantitative testing procedure was developed in the process in order to meet the specific need of testing measures of disability and health state suitable for international comparisons.

Information was collected on the question response process, patterns of interpretation, and evaluation of decision-making patterns. This information was then used to help identify potential response error and to test the suitability of the questions for the purpose for which they were designed to generate a meaningful, internationally comparable general prevalence measure for disability. The pattern analysis was particularly advantageous. Examining consistencies and inconsistencies across various questions allowed for an evaluation of the Washington Group questions without establishing existing questions as a gold standard. This pattern analysis allowed for old and new questions to be compared while maintaining a neutral or agnostic view of the other questions.

The results indicate the usefulness of this approach for testing the design of cross-national indicators, as well as lending support to the reliability of the particular measures developed (and now adopted) by the Washington Group on Disability Statistics.

DISCUSSION

The discussion focused on three issues: vignettes, item pools and CAT, and the Washington Group.

Vignettes

Several participants asked questions: Are some of the differences in reporting on work disability in the Netherlands and the United States explained by differences in the interview context and the organizations conducting the surveys? Can some of the differences be explained through the framing effect, that is, the location of the questions in the questionnaire? Can some of the results showing that African Americans have disability profiles at ages 10 years younger than whites be explained using vignette methodology? Has the vignette methodology been used to look at racial or ethnic differences within the United States?

Arthur van Soest responded that the surveys were Internet surveys in both countries. However, he noted that they were not exactly the same complete survey, so there is possibly a framing effect. One of the interpretations of what the authors found is that the country's, and its institutions', specific context does make a difference. The researchers also experimented with telephone interviews, but reading all those hypothetical stories did not work as well. One kind of experiment could be to ask somebody in the Netherlands about a hypothetical person in the United States and vice versa. However, they thought that would be too confusing and so did not use that approach.

He noted that in the Netherlands and U.S. study, race was not included, mainly because in the Netherlands there is not enough variation in the data; items such as education, gender, and age were included. In response to the question about whether the vignette methodology has been used to look at racial or ethnic differences in the United States, van Soest replied that it could be used in such surveys as the Health and Retirement Study, but he does not know if it has been used. The focus for using vignettes to date has been almost exclusively on cross-national comparisons.

Item Pools and Computer-Adaptive Testing

Noting that the vignette scheme is particularly suited to identifying and making adjustments for DIF, Robert Hauser wondered whether an item pool for CAT would be valid if it had DIF in the items. To what extent have item pools for CAT been tested for DIF?

Whether an item bank has DIF is based on where one is looking for it. There are many different levels at which one can look. The small number of items that show differential functioning can be removed from an item bank, especially if the bank has hundreds of items. Also, one can calibrate a group-specific item differential so that, for example, it would have a different item differential for Dutch and U.S. populations. It would be very similar to the vignette scheme.

It was noted that application of these methods across different populations groups would be important. Gender is an example. The standard general health questions do not show any gender differences, but adjustments using these methods might show that there is a gender difference and that gender accounts for the difference between self-reported symptoms, for which there are always gender differences in the general health.

The Washington Group

A question was asked if there are some domains that the group simply cannot get to work across the multiple sites being evaluated.

Julie Weeks responded that, anticipating that some domains would be harder than others to work with across the multiple sites being evaluated, the group started with some of the easier domains. In the Budapest Initiative, they are encountering difficulty with two domains—how to ask about pain and fatigue cross-culturally. How people interpret the concepts of pain and fatigue and how much they are willing to admit to having them are very different. Those two domains are in the third round of testing, without success as yet. Obviously, this work needs to continue.

Connie Citro (Committee on National Statistics, DBASSE) commended the Washington Group initiative, which is clearly going back to basics as the way of making a start at getting some very carefully tested questions that will provide basic monitoring information across a whole range of countries. She said that is very important work.

5

Measuring Functioning and Disability in Context

This chapter focuses on a number of cutting-edge areas in conceptualizing factors external to individuals that are key either to identifying changes in functioning or to modifying experienced function, and the context in which people function, and the need to measure them in context, taking into account suitability for population surveys and relevance for monitoring trends. The presentations covered three issues:

1. Environmental barriers and modifications
2. Behavioral adaptations
3. The utility of participation measures in population surveys

INCORPORATING ASSISTIVE TECHNOLOGY AND HOME MODIFICATION MEASURES

The objectives of Emily Agree's (Johns Hopkins Bloomberg School of Public Health) presentation were to discuss how accommodations, like the use of assistive technology or the use of home modifications, relate to individuals' functioning and to describe findings from a pilot study on assistive technology, fielded by Agree and colleagues in 2005, designed to develop survey questions on assistive technology and home environments.

The use of assistive technology among older adults has increased in recent decades, especially for mobility and bathing. On the basis of analysis of several national data sets, an estimated 14–20 percent of people use some kind of assistive technology, regardless of difficulty with tasks. Use of assistive technology may have contributed substantially to declines in dependence on personal care over time. Estimates show that in the 1990s nearly half of Americans aged 65 and older had home modifications or adaptations to reduce barriers. Results from the 2006 Health and Retirement Study (HRS) showed almost 70 percent of people aged 50 and older had a home modification. In general, both clinical and population-based studies suggest that the use of assistive technology and home modifications may improve functioning and quality of life, expand participation in activities, expand neighborhood mobility, and protect caregivers' health.

Disability can be conceived as a gap between individuals' capacities (physical, cognitive, and sensory ability) and their performance in daily activities and participation in social life. The ability of individuals to translate their intrinsic capacity into successful performance is affected by the context in which they perform each of these activities. What is required in order to do an activity depends on the specific task—getting out of bed, socializing with family and friends, or going to work. Each of these activity demands, in addition, has an activity-specific environmental context, which can incorporate barriers to accomplishing that activity as well as dimensions of the environment that actually facilitate conducting that activity.

The extent to which individuals can translate capacity into performance also depends on what they themselves can do to change the environment or to change the demands of the activity by adapting or accommodating, what has sometimes been termed compensation or using a compensatory strategy. Compensation includes the use of human help (both formal and informal care), changes in the way tasks are done (including the use of technology), and changes that are made to the home environment.

To expand on the concept a bit, if disability is an activity-specific gap between individual capacity and performance, technology expands the capacity of the individual and environmental modifications reduce barriers in the environment.

Current Survey Measurement of Assistive Technology Use

A review of six major national surveys—(1) HRS, (2) the National Health Interview Survey (NHIS), (3) the National Long-Term Care Survey (NLTCS), (4) the Medicare Current Beneficiary Survey, (5) the Medical Expenditure Panel Survey, and (6) the Survey of Income and Participation—shows that in the past few years there has been a proliferation in the number of surveys including questions about assistive technology. However, terminology varies across surveys (aids, special equipment, adaptive devices, medical devices or supplies, etc.), and so there has been little agreement in the data.

The level of detail on device use also varies across surveys. Questions may be asked globally about the use of devices for all activities or specific to each task. Similarly, detail on the devices themselves varies, whether a general question about overall use is asked about special equipment or whether individual devices are catalogued.

Another area of variation is the characterization of use. Particularly important has been the use of a reference time period in describing use. Basically, only two surveys have actually used any reference period at all: NHIS asks if people used devices now, and NLTCS asks about devices used in the past week.

Finally, these items most often are still embedded in ADL or sensory limitation questions, most often restricted to those reporting any difficulty. This approach uses ADL questions as a screener for questions about health and technology use. Consequently, if people have successfully resolved their difficulties with a task by using some adaptation, they would not be screened into the assistive technology questions at all; in national surveys, these people are being missed.

With regard to the environment, in general there are very few questions about the home environment in national surveys of health and aging, and surveys generally do not distinguish between the existence, addition, and use of adaptations or devices. Therefore, modifications are often only asked in terms of use because they are asked in the same way as other assistive technology questions, whereas dwelling features are asked without reference to disability.

There are a series of instruments used in the rehabilitation profession to do home assessments, which can involve a long checklist of classifications of potential barriers in the home. Their suitability for national population surveys is questionable because they are very long and use subjective terminology. It is hard to figure out how to objectively use those checklists at a population level (rather than at a clinical level) to assess the features of home environments and their potential to impose barriers.

Agree noted some of the limitations of current measurement approaches. Surveys of health and aging often conflate the use of assistive technology with disability by embedding questions on device use in disability questions and restricting device questions to those who report difficulties. Assistive technology tends to be task specific by design, and so survey questions on assistive technology should be task specific; however, they need to be separate from questions about task difficulty. Such an approach would help people understand how effective assistive technology may be in reducing difficulty *with tasks*.

Another limitation of existing approaches is that the current assistive technology measures are quite basic. They are often broad dichotomies that do not capture patterns of use, such as "Do you use a cane?" The response, yes or no, does not provide information about the frequency with which a device is used, the location, and other aspects of use that give a more nuanced view about whether people have tradeoffs between devices, use them in different places, or use them in different ways.

Finally, the home environment is even more neglected in surveys than assistive technology. There are very few measures of features of the home environment, and they do not distinguish whether the respondent added a feature, whether that person uses it, or whether someone else in the household uses it. No attempt has been made to translate some of the clinical assessment tools to national survey questions and test them. Of particular interest is to better understand how and when environmental modifications are made, relative to the progression of a disability, the nature of the use of modifications, and whether they address the barriers in the home environment.

Pilot Study on Assistive Technology

In 2005, with support from the Office of the Assistant Secretary for Planning and Evaluation in the U.S. Department of Health and Human Services, Agree and her colleagues conducted a nationwide pilot study to develop and test survey instruments on assistive technology use and the home environment for national surveys on health and aging. The goals of the pilot study were to design questions that are useful for people of all abilities, use positive language, and assess device use across environments and activities.

The study included 360 persons aged 50 and older of all levels of ability. The sample was racially and geographically diverse and oversampled persons in assisted living facilities. The sample was not representative and there was no conversion for nonresponse. However, the sample for the pilot was weighted to match the 2005 NHIS to provide somewhat representative estimates.

Roughly equal numbers of people in age groups 50–64, 65–79, and 80 and older were included in the sample in order to be able to test items across a wide age spectrum. The need for and the use of assistive technology vary widely between the younger end of the age range (those who are still working, in good health, and familiar with available technologies) and the oldest old (who tend to be retired, experiencing more dynamic health changes, and less familiar with available technologies).

Disentangling assistive technology use from ADL difficulty meant setting up two types of questions: first, establishing task-specific device use without regard to difficulty, and, second, establishing "difficulty" in activities when device use is taken into account. Agree and her colleagues referred to this measure as "independent functioning."

First, data on device use were collected using a specific reference period and asking about frequency by task and location. For example, for mobility devices, the person was asked: "In the last 30 days, have you used a cane, [walker, wheelchair, and scooter], yes or no?" Respondents who said yes were then asked a series of questions about how often they used their device for relevant activities, such as getting around inside their home or building, transferring between home and outside, and getting around outside their home or building. They were then asked about the frequency of use: "In the last thirty days, when you got out of a bed or chair, how often did you use your cane to help? Would you say every time, most times, sometimes, rarely or never?"

The approach for the home environment questions was to inquire about the existence, acquisition, and use of adaptations or devices. The existence of home modifications is important for aging in place and the potential adaptability of home environments. Acquisition is important to ascertain for home modifications (not for portable assistive technology) because the item may not have been added by (or for) the respondent, even if it is clearly an adaptation for disability. The most obvious example is grab bars, which may be a part of general bathroom renovations but may also include any accessibility modifications that are often included in new construction to meet requirements of the Americans with Disabilities Act. Or a feature may have been installed for another current or previous household member. It is also important to collect information on the frequency of use for these items and to limit use to a specific period of time, say the last 30 days, because some items may be installed for safety or used only at certain times. For example, it is very common to use a portable commode at night only.

The second part of disentangling difficulty from device use was to assess "independent functioning," a measure of task performance that incorporates use of devices without human help. Agree said she and her colleagues suggest refining ADL measures to ask about activities in a way that represents independent functioning. These items differ from those commonly found in national surveys in two ways. First, they focus on the level of difficulty with activities when using assistive devices and without help from another person. Second, the items are tailored to mention a specific list of devices and features that are reported by each respondent in separate questions about device use and use of home modifications.

Items were designed to determine how well the person can do a task using the particular devices without help, and so some modification was needed to allow respondents to volunteer that they never do the activities without human help. Respondents who said that they never do the activity without help were then asked "using your [device(s)] could you do this task by yourself?" Almost all respondents who answered that that they "never do the activity without help" responded to the follow-up question that they "could do" the activity. The follow-up question was therefore eliminated and respondents recoded as having severe difficulty with the task.

There were concerns about whether respondents would understand the concept of "difficulty with assistive technology and without help" and be able to answer such questions. To examine that issue, traditional functional limitation questions also were asked in the pilot, and Westat conducted behavior coding on a substantial portion of interviews. The responses indicated that, particularly for the ADL questions, they performed quite well. They required fewer clarifications and less probing than standard Nagi functional limitation questions, and they had a very small percentage of respondents (0.05 to 1.0 percent) who gave inappropriate answers (such as "don't know" or "refused to answer"). They also scaled very well.

Finally, the items can be used in ways that help on a policy level to target potential groups who may need home modifications or help. About 15 percent of adults age 50 and older in the sample who had severe lower body limitations also had at least one unmodified barrier in their homes that was either in the entry to the home, inside the home, or in the bath area. The percentage who could benefit from an environmental modification varies by location in the home. For example, 9.1 percent of respondents had a severe lower body limitation, must use at least one step to leave home, and have no railings or ramp at the entrance; 7.0 percent had a severe limitation and no separate shower, grab bar, or seat in the tub. However, only a small percentage of adults aged 50 and older (2.8 percent) had severe body limitations, living space on multiple floors, and no stair glide. About one in five older adults (20 percent) had a severe lower body limitation but no safety features (grab bars or raised toilet seat) for the toilet area. Overall, nearly one in four adults aged 50 and older (23 percent) could be candidates for environmental modifications in their homes. Broadening the criteria to include anyone with a lower body limitation (irrespective of severity) results in a much larger group—up to 43 percent of adults aged 50 and older. These data show that there is a substantial group of people who have functional limitations, as they were measured in the pilot study, and who face barriers in their homes. Thus, these data can be used to determine the potential need for targeting effective interventions.

INCORPORATING QUESTIONS ON BEHAVIORAL ADAPTATIONS IN FUNCTIONAL LIMITATION MEASURES

Carlos Weiss (Johns Hopkins University School of Medicine) reported on some of the work in the area of behavioral adaptation as a way of trying to improve population health. He offered some thoughts about how future studies might be able to design a set of meaningful measures for population surveys.

One of the main, and important, reasons that behavioral adaptations matter is prevention of more advanced disability and the identification of people who could be targeted for effective intervention. Participation occurs on a continuum, from having difficulty doing tasks to restricted participation to dependence, and behavioral adaptations occur at many different levels along that continuum. However, it may make sense to first focus on understanding behavioral adaptation by studying it in the absence of dependence, and even in the absence of difficulty, to understand some of its salient features. Although behavioral adaptation certainly occurs in the presence of difficulty and even dependence, it may play a more salient role in the absence of these, as an early change that precedes difficulty. It follows, however, from this preventive focus, that behavioral adaptations in the absence of difficulty may be less tightly linked to such outcomes as institutionalization and death than more advanced manifestations of dependence. That is a way of saying that if research focuses on behavioral adaptations alone, one is looking at a group of people who may be at lower risk for some of the important disability outcomes.

Data compiled by Thomas Gill and colleagues (1998) on the relation between different stages on the continuum and important endpoints, such as admission to a nursing facility or dying within 3 years, illustrate the fact that focusing on behavioral adaptations in the absence of dependence, and even difficulty, means that one is looking at a group with lower risk of the more downstream outcomes that are necessary anchors for disability research. However, there is strong public health imperative to understand this part of the continuum. Studies of the distribution of the Medicare population have shown that people with dependence are a small but important minority (only about 4.0 percent), and people who are independent but with difficulty are also a minority (about 4.5 percent). People without difficulty, many of whom are using behavioral adaptations, are a majority (about 50–55 percent) (Shumway-Cook et al., 2005).

In addition, disability fluctuates. Sometimes it lurches catastrophically toward a bad outcome; more generally, over long periods, it tends to be gradual and to progressively diminish participation. This phenomenon occurs in older adults and also for some younger adults who have significant chronic disease. These facts are justification to use behavioral adaptations to attempt prevention, but they also mean that by this very framework some adaptations will no longer be needed, and they are bound to have less predictive accuracy. The main question, then, is whether one can identify a modifiable preclinical phase at which early intervention is more effective than waiting for more overt presentation. The emerging answer to this question is quite probably yes.

What are behavioral adaptations? Weiss noted that it is useful to think of behavioral adaptations within a compensatory strategy framework, taking a slightly different perspective than a focus on dependence and difficulty. When human help replaces the person actually doing the task, perhaps that is no longer a compensatory strategy; rather it is something different. However, there are different degrees of human help, and so behavioral adaptations often occur with, and without, the concomitant use of other strategies. A useful way to narrow the otherwise innumerable differences in the ways people participate in important social roles is to focus on behavioral adaptations that are responses to mild to moderate impairment, to ensure that the behavioral adaptations are occurring despite normal conditions and to ask whether they involve performing tasks or roles in a way that is not usual for that person. Consider walking as a prime example: one may sometimes walk more slowly. That, by itself, is not a compensatory strategy, unless the person recently injured a tendon in the foot. If the slow walking is under normal conditions, rather than on a rocky trail or on a slippery surface, it may be more of a compensatory strategy in the form of behavioral adaptation.

The literature on behavioral adaptation has already shown that preclinical disability, defined as task modification in the absence of difficulty, is a strong risk marker for the development of difficulty (Fried et al., 2000). Preclinical disability measurement is reliable and has both construct validity and predictive validity (Weiss et al., 2007). Weiss presented the results of some recent analyses, involving subsequent questions and with outcomes going further out in time. The data come from the Women's Health and Aging Study II, which comprised 436 women aged 70–79, selected from among the two-thirds most high-functioning women living in the community. Framing may be important in this work, because the questions on task modification—that is, preclinical disability—came after the standard questions about difficulty for a health or physical reason. Interestingly, by far the most common behavioral adaptation for walking among this highly functional group was to slow down. This notion is supported by a large nonsurvey literature made up of physiological and experimental studies with small numbers of healthy older adults or older adults with specific conditions, such as diabetic neuropathy.

Using data from the Women's Health and Aging Study II, Weiss and his colleagues examined 3-year outcomes among women who were initially walking at least eight blocks outdoors a week. They modeled the probability that the women would cease to walk as much according to baseline level of self-selected walking speed, whether the women had a physical impairment in strength or balance, and whether they started to adapt the way they walked at some point after the baseline measurement. Self-report of starting-to-adapt behavior (after baseline) was the variable that spread the risk the most, much more so than baseline impairment. The self-reported behavioral adaptation data complemented the data on walking speed. There were follow-up questions in this study that asked the women why they had started to adapt—why they changed the way they walk. The respondents were shown about 20 symptoms related to health conditions as response options. For the tasks of walking half a mile and climbing 10 steps, pain and low energy emerged as significant reported causes; for climbing 10 steps, safety was also an important issue.

In summary, these results show that compensatory strategies are the result of a complex interaction between demand and capability. They are a manifestation of diminished reserve relative to demand. A compensatory strategy can take at least two forms in such a scheme: offloading demand or increasing capability through augmenting leverage. In other words, a compensatory strategy can mean doing less or staying active despite an impairment. Compensatory strategies may also be revealing about additional factors, such as the environment, but one of the things they reveal is about health status and physical capacity, suggesting that there is a lot of qualitative heterogeneity among different compensatory strategies. For example, among women who reported that they changed the way they walk, there might be one type of compensation that essentially involves doing the same amount of walking—a leveraging strategy. These women are staying as active as they were before, in different ways. They could walk slower and spend more time walking. They could do other things that involve interacting with the environment and other human beings and using devices. They could seek good light and concentrate on walking, or they might make sure to have a walking cane and walk with someone who is not too talkative and does not interrupt concentration on walking. At different times, a compensatory strategy could instead involve cutting back on the demand of that task, to things that involve walking slowly. These different types of compensatory strategies would have different outcomes in terms of the amount of activity performed or time spent doing the activity.

Weiss pointed out that the study of behavioral adaptation is already taking place under many different names in population surveys, such as subclinical disability, time use, avoidance, and life space. Some of the lessons learned include that it is possible to identify early adaptations that appear to have meaning for the purpose of prevention, and that self-reports complement objectively measured performance. In particular, the heterogeneity among adaptations—staying active versus doing less—correlates with time-use ways of looking at behavioral adaptations. One of the exciting things about the time-use data is that they involve an external scale, in contrast to the internal scale used for adaptation.

Behavioral adaptations are an area worthy of further exploration. Research questions that that should be addressed in the near future include

(a) whether one should allow the high specificity and low sensitivity of a longer, more ambiguous time frame for some of these adaptations; (b) a more refined understanding of the reasons for changes, that is, why people change how they perform these activities, including personal and environmental factors; and (c) which of these behavioral adaptations are modifiable and therefore high-priority targets for intervention.

UTILITY OF PARTICIPATION MEASURES IN SURVEYS

Gale Whiteneck's (Craig Hospital, Englewood, Colorado) presentation focused on the relevance of measured participation from his perspective as a disability and rehabilitation researcher. His presentation addressed five issues: what is participation, why is it important, how is it measured, why is it hard to measure, and how should it be used in population surveys.

What Is Participation?

The World Health Organization's International Classification of Function, Disability and Health (ICF) model focuses on participation and defines participation as involvement in life situations. However, this is not a good definition, because it does not adequately differentiate activity from participation. Participation is performance at the societal level; it is fulfilling the social roles of being a worker, a volunteer, a homemaker, a spouse, a friend, a grandparent, a citizen, a neighbor. It is being an active, productive member of society who is well integrated into family and community life.

Why Is Participation Important?

Participation is a major construct in all of the disability models. Full participation in society is the goal of the Americans with Disabilities Act. It is the ultimate goal of rehabilitation. It is what people with disabilities and their families are most interested in—they are interested in functioning in life. They are more interested in fulfilling social roles than whether they can dress or ambulate well. Although the role may change, it is no less important in late life—it is what makes life worth living.

How Is Participation Measured?

In comparison with activity limitation (ADL limitations) measures, participation restriction measurement or participation measurement is more recent in the field, it is measured less frequently, and it is less well developed. There is no gold standard or agreed measure of participation. Participation is often measured with time-use methods, as described in an earlier session of the workshop.

Domains of participation are evaluated with common examples asking self-reported frequency or counts using stylized questions. If the interest is in the domain of productivity, the hours per typical week spent working, homemaking, in school, or volunteering are often asked. For the social domain, information would be obtained on counts of friends, frequency of contacts with family, involvement in a romantic relationship, and whether married and living with a spouse. For the community domain, questions may be asked about the days per week outside the house, times per month shopping, eating out, going to church, or going to the movies, or the number of times involved with community organizations.

ICF provides this taxonomy or classification of participation in the chapters on activity and participation in a list of ways people participate in interpersonal relationships, major life areas, and community, social, and civic life. However, ICF does not provide a measurement of participation, only a classification.

The Craig Handicap Assessment and Reporting Technique (CHART) is an early participation measure developed at Craig Hospital (Whiteneck et al., 1992). It sums across relatively objective items in domains weighted by importance of the item as perceived by the general population, and it computes a participation score within domains for people with disability in comparison to the norms of people without disability.

Why Is It Hard to Measure?

There are individual preferences about how to participate, and choices in communities, but there is no list of participation items that applies to all. For example, not everyone has to work to participate in society; not everyone has to be a full-time homemaker or be married to participate in society.

People with disabilities do not want to be judged on the basis of norms for people without disabilities: CHART and similar tools have been criticized on those grounds. Therefore, there has been a recent focus on more subjective aspects of participation. The Participation Objective, Participation Subjective measure, which is part of the measure known as Living Life After Traumatic Brain Injury (Brown et al., 2004), is a good example of that type of measure. The objective section measures frequency or counts of involvement in various elements of participation. The subjective section asks how important each item is to the individual and how satisfied the individual is in each area or item. For example, a person is asked how many hours the person works, how important is working to the person, and if the person is satisfied with the amount of time spent working. The results are performance and satisfaction scores weighted by importance.

Other metrics are being tried. The Participation Measure for Post Acute Care focuses on the difficulty of participation more than the frequency or quantity, although some of those latter items are also included. The community participation indicators (CPI) include enfranchisement items measuring the extent to which people feel engaged, accepted, and valued in their communities.

There are many reasons why it is hard to measure participation. The choice among ways of participating means there really is no hierarchy of participation. Ways of participation cannot be arrayed on either a difficulty or a value continuum, just a frequency continuum. There certainly is no expectation that one has to do each piece or that there is a hierarchy. For example, it is not comparable to the situation in which a person stands before walking and then running, so that anybody who can run can be assumed to be able to walk and stand. If one is talking about work, homemaking, and being a student, which is the most difficult? Assuming that being a student is easiest and homemaking is most difficult, one cannot assume that a person who is a full-time homemaker is also a full-time worker and also a full-time student.

The above illustrates that there is no hierarchy within the elements of participation. Without hierarchy, the assumptions of item response theory, which is the current state of the art in developing measurement tools, at least in the participation arena, are not met and may not be appropriate.

How Should Participation Measures Be Used in Population Surveys?

Whiteneck argued that participation measures should not be used to expand the definition of who is disabled. Measures of activity limitations should be used to distinguish people with and without disability, as well as among types and severity levels of disability. Once the disability measures have been selected, participation measures can be used to compare the extent of participation among groups of people with and without disabilities or among people with disabilities of various types and severities.

Population surveys can be used to assess the integration or equalization of opportunity for a population of concern—people with disability in this case—compared to the general population. Participation measures can be used in population surveys to assess the participation gap between people with and without disabilities, trends over time, and the effects of environmental interventions. For example, participation measures developed in rehabilitation, such as CHART and CPI, have been used in Colorado, piggybacking on the Behavioral Risk Factors Surveillance System of the Centers for Disease Control and Prevention. By using these measures in general population surveys, people with and without disabilities can be compared in terms of their responses.

Another example of using participation measures in population surveys is the Canadian Participation and Activity Limitation Survey. That is a quinquennial postcensal survey, targeting people with positive responses to two disability screeners in the census. In 2006, the sample was 47,500 persons. It is a lengthy branching survey primarily focusing on activity limitations, but it also includes participation and quality of life.

To conclude, participation measures should be used to determine the extent to which people with disabilities are fully participating in society.

DISCUSSION

Participants asked several questions for clarification or elaboration mostly focused on three topics: whether there is an assumption of decline in measuring behavior modifications, participation measures, and assistive devices.

Whether There Is an Assumption of Decline in Measuring Behavior Modifications

A comment was made that people not only change their behavior, say in walking, but also take on other activities. There is a substitution of behavior, and the choice element needs to be included. Carlos Weiss responded that decline should not be assumed. There may be some sort of a general tendency for decline among people who have chronic diseases, but it should not be assumed. He referred to the work done in the 1980s by Lois Verbrugge on diaries of changes in health status and change in function, realizing that there are some people who are remarkably stable over time, and some who are able to change a little, maybe through the use of different strategies, but then maintain their activity levels over a long period of time. It was noted that activities such as mobility are essential activities that are the stepping stones to other more volitional and complex activities. People do have to adapt for the set of basic activities to function on a daily basis.

Participation Measures

Participants raised two questions: Is it appropriate to be looking for participation measures that work across different population groups, such as clinical population surveys of the general older population? Can universal measures of participation be developed for the older people who may have progressively declining function?

The concept of participation is life-long: participation as children, participation as adults, and participation as late-life adults. Age-appropriate norms based on different items for participation with enough options are required, said one participant. An obstacle for older adults is the real dearth of opportunities for meaningful roles. What is needed is a clear measure of participation and a clear measure of environments in order to know how the two interact. If enough opportunities for meaningful roles do not exist, that is a problem of the environment and not of the person. So interventions to offer meaningful roles in late life can be introduced to be able to measure increased participation. However, a lot of variability exists, both among individuals and in terms of environmental demands. Just extending the list of participation measures as a way to measure and understand disability may not hold either across age groups or across different culture groups and subgroups in the population.

The difference between these participation measures and the more usual measures of activity limitations has to do with the fact that people value what they participate in differently. It is going to hinge on asking people how important each of these areas is to them so that each person has his or her own valuation scale. So it is not what you are doing or how often you are doing it, but whether it relates to what you value. Perhaps across cultures and across age groups, people value different things.

Simply drawing or producing correlations between disability states and levels of participation in late life does not give a good specification of what is going on. People have lifelong patterns of participation, and what people do affects other people and has health consequences, including disabilities. Longitudinal data are needed to make sense of such relationships.

Assistive Devices

In response to a question, it was noted that training is needed to use an assistive device, and the process of adopting a device may be complicated. People do not land on the right device the first time; they try the device, adjust it, or substitute it for another device. There is the problem of abandonment of a device; is it replaced, are new devices added, or are they used differently? These possible adaptations are hard to measure on a national survey instrument, and they were not addressed in the pilot study described by Agree. Nevertheless, many participants said it is an extremely important area to develop.

Other issues were raised. There is the problem of the risk of injury in using a device. This is particularly true in wheelchair use, and it also affects the people pushing wheelchairs. Training is important to ensure that people use a device appropriately. The intersection between the device and the home environment needs to be addressed. Older people function in many environments, but researchers tend to concentrate on the home environment because of a focus on the most impaired subgroup of the older population, people with the greatest problems in the most basic activities. However, one needs to understand the environment in terms of the areas outside the home, the neighborhood environment, and the environments that are specific to those activities that are of value to people, including how they get to those activities—the transportation and communication environments. This area also was not addressed in the pilot study. It may be far too simplistic to say that environmental barriers impede participation and environmental facilitators enhance participation.

6

Research and Development Toward Improved Measures of Late-Life Disability

In the last session of the workshop, participants highlighted their perceptions of some of the research topics and priorities that emerged from the presentations and discussions, regarding important and potentially fruitful areas for research and development, given the goal of improving the measurement of late-life disability in population surveys. The session began with a panel of three workshop presenters who led off the discussion, sharing their initial thoughts on areas for research and development, followed by a general discussion and comments from all participants.

PANEL COMMENTS

Thomas Gill identified two areas of research opportunities applying methodologies discussed during the workshop. The first area was the application of time-use data to study disability, but in terms of recovery from disability. One cannot have a full and complete understanding of disability unless one understands the recovery process, he said. It is increasingly clear that disability is a highly dynamic process. There are opportunities that have not been fully realized in terms of an expanded array of functional outcome measures with regard to recovery. That is where time-use data could provide some new and unique insights, potentially for shorter periods of time than might otherwise be possible in the context of the traditional disability surveys, because the recovery process typically occurs over a relatively short period of time. Time-use data could prove quite valuable in providing a more complete understanding of how and to what extent individuals who become disabled, say by a hip fracture or other injury or any hospitalization, regain the full array of their predisability function.

The second area is the use of vignettes, a promising method that calls for further research and experimentation. The focus to date has been on cross-national comparisons. Can vignettes be used within the United States to better understand the differences in disability among different ethnic or racial groups or between genders? For some of the disability patterns and disparities described in the workshop, Gill noted that the African American population has a disability profile that mirrors the white population, but for 10 years younger. Could some of that difference be understood through the use of vignettes in identifying the different pathways to disability and distinct populations in the United States?

Arthur van Soest highlighted four areas. First he asked: What is the definition of late life? This workshop is about late-life disability. One can distinguish between the older old (say, 70 years and older) and the younger old (say, ages 50–70). There are differences in the issues concerning these two groups. For the older group, such things as home adjustments, devices, informal and formal care, and neighborhood characteristics seem to be most relevant. For the younger group, the relationship between work and disability seems crucial, including working conditions, employer attitudes and accommodations, job characteristics, stressful and physically demanding work, and stress due to unemployment.

Van Soest noted that most presentations during the workshop focused on the older group, but the younger group also deserves attention for two reasons. First, a substantial percentage in this group already suffers from disabilities; second, health and disability, as well as employment characteristics, at younger ages, are predictive of disabilities at an older age. So the issues may be different for the two groups, but both are worth studying. One of the implications is that longitudinal data are needed to follow people over a long period of time. He noted that the Health and Retirement Study is now reaching the stage at which one can start to study these long-term effects with good longitudinal data.

Second, van Soest noted that health is well known to be related to socioeconomic status, and the same is true for disability. Whichever method is used to measure disability, the data will show a negative association between prevalence of disability and socioeconomic status—education level and income or wealth. Determining the causal mechanisms that may lead to this association is an important research question.

He said that the workshop pointed to several potential mechanisms that may explain a causal effect of socioeconomic status on disability prevalence: working conditions (physically demanding work, stress), differences in health behavior (e.g., obesity is more prevalent among less educated people), as well as the affordability of health care or home modifications and devices. This is also an issue for which international comparisons will be useful, since the extent to which home modifications and special devices are covered by insurance or subsidized for low-income groups varies across countries.

The third area van Soest identified is disability and quality of life. The idea of studying time use in relation to disability is very interesting, particularly in how people cope with their disabilities and how disability affects the quality of their lives—not only the effects on time use, but also on social participation, a person's economic situation, and, ultimately, well-being and happiness.

The fourth area concerns survey methodology. Many presentations identified existing problems, although often no one knows how to solve these problems. Van Soest noted four issues:

1. Which questions can we ask and how can we ask them? For example, seemingly simple questions apparently lead to inconsistent answers from husbands and wives.
2. Some things can be determined better in different ways. Using biomarkers is a promising development that is now feasible. In addition, using administrative data—on income, wealth, pension entitlements, and perhaps also on health care use, as some European countries do—will help to improve data quality. The new approaches are not suggested to replace survey data, but to supplement them.
3. In terms of sample design and sampling frame, sometimes a representative sample of the population is not needed, but often it is. In that case, the selection problems seem salient: people with a disability will be less likely to participate and more likely to drop out. Proxy interviews offer a partial solution. Although there may be problems with proxies, that does not mean we should not use them. More research on the quality of proxy interviews is warranted.
4. Internet interviews are a promising and cost-effective way for the future but obviously suffer from selection problems even more than other approaches. The use of proxies in Internet interviews seems something that is worthwhile to test.

David Weir spoke to the issues of definition and measurement. He said he was sure most of the participants had been struck by the box diagram presentations on what disability is. The concept of disability is complex because it is relational. The key elements are health, the physical and mental capacities or abilities of an individual; the environment, much of which is completely beyond the individual's control; and then, in the middle, the behaviors and psychological adaptations through which people translate their abilities into activities and rewards they are able to achieve.

What should surveys do in light of this conceptualization? The nationally representative population surveys, first and foremost, have to be very clear about the measurement of health and about the measurement of abilities and capacities. Weir noted that the description of the Washington Group's effort was great. That kind of rigor in developing self-report questions will allow conduct of comparable studies across population groups.

Clearly, there is a place for performance measures, which are, by their nature, attempts to measure capacities and abilities. The idea is to abstract as much as possible from the environment so that a person with one set of capacities might be disabled in one environment but not in another. However, information about a person's health is also important, in part because if one is focused on late-life disability, one will want to know about aging and how that translates into life experience. Weir said that an operating definition of aging is a decline of physical and mental abilities and capacities, and no one has found a solution to that. Researchers want to know about what is being modified—whether medical intervention or people's behavior or something is changing the pace at which people's capacities decline over time. The only way to have confidence in those types of findings is to have solid comparable measures of those capacities and abilities in surveys. So that should be an important priority.

Given the nature of declining function with age, any survey is going to need to rely on proxy reporting to capture some of the people who have a disability. If people who cannot respond for themselves are not represented in a survey, it will miss what is going on. So there is a need for measures that are interchangeable between proxies and self-reports. Weir noted that one area that was not discussed much during the workshop is whether the existing health measures cover all of the domains of health that are relevant to disability in the sense of limitations in activity. Turning next to the other end of the spectrum, the environment, surveys of people can address what is going on in the environment, and one can also survey environments. That is actually being done; for example, looking at how transportation systems function, at whether people have access to places of work, and at shopping malls.

Finally, work on adaptation is the research frontier. Surveys have a role to play in the sense that people can be queried about whether they are using assistive devices and how they are adapting. They can also be asked how they feel about their health and how they feel about the activities they are able, or unable, to engage in. These types of inquiries have to be viewed as more of a research activity for which nationally representative population surveys are not necessarily required because they would be for monitoring trends and evaluating health policies.

DISCUSSION

During the general discussion, participants identified several areas for further research and development that may lead to improved or enhanced measurement of late-life disability. Among the several comments and questions, six subject areas surfaced as the focus of most of the discussion: (1) clinical classification of disability, (2) measuring levels and changes in the levels, (3) vignettes, (4) inclusion of the working-age population, (5) the purpose of surveys, and (6) measures of participation and the environment.

Clinical Classification of Disability

Bob Wallace (University of Iowa) said that all the areas for further research identified so far are important, but he did not think that disabling illnesses of older people are being characterized very well. A good deal of surgery, as well as many functional disorders that affect the disability, such as musculoskeletal disability, are not being talked about or measured. A commodity called disability, even if it is perfectly measured, really provides little information. For example, neuropathy is different from constipation, which is different from old fractures and old football injuries. Better clinical classification of disability is needed of the kinds of dysfunctions and disabilities of older people. Only with such classification will it be possible to move to the next level, which is prevention. He observed that there are ways to prevent various kinds of disabilities, and they should be assessed in population surveys.

Wallace also noted that late-life disability should be distinguished from disabilities that people have had most of their lives—these people are different, and their trajectories are different. It was noted that because of sustaining medical care, for example, there will be more older people with conditions such as Down syndrome and cystic fibrosis than before.

Jack Guralnik observed that although significant advances have been made moving forward from the medical model of disability to assessing function and disability in the whole older person, to understand trends in disability and to consider interventions to prevent disability, researchers will need to start looking more at the diseases underlying disability. These types of data would be very valuable for determining intervention strategies to prevent disability. He noted that this topic was not discussed at the workshop. Some surveys have included a question on the main cause of disability, but no one has a good sense of the validity of the responses. There have been statistical analyses looking at the association of disease and disability, but it is hard to know if it shows just a cross-sectional association or is causal. Even longitudinal studies have some problems. There is a tremendous amount of comorbidity in the older population and it is hard to sort out the main cause of disability.

Difference Between Measuring Levels and Measuring a Change in That Level

John Haaga (National Institute on Aging) observed that information on change in the level of functioning is more important and a better prognostic indicator of potential disability for an individual than just information on levels. This information is useful in clinical work, and it can be measured even one time in a survey. For example, there is no right amount of time to sleep, and the number of hours someone sleeps is not a good measure of anything other than sleep, but a recent change in the number of hours of sleep is a warning sign. Similarly, gait speed and other items that are good measures of disability become even better measures or better prognostic indicators if there has been a recent change in the item (in either direction).

Vignettes

Richard Suzman (National Institute on Aging) observed that vignette methodology remains a promising experimental technique that is worth much more experimentation and that much of the success is going to come from writing good vignettes. Perhaps presenting them in different ways and ensuring that the respondents have understood the vignette when they give answers to the questions will enhance comparability of self-reports across important population groups.

Including the Working-Age Population

Andrew Houtenville (New Editions Consulting) stressed the need to include younger ages in population-based surveys. Researchers working with disabilities among the working-age population would benefit from help in terms of defining disability either in population-based surveys or for programs such as Social Security Disability Insurance and Supplemental Security Income benefits. With the increase in the age for receiving Social Security benefits and other economic forces, work is going to be an increasing part of older life.

The work done with vignettes focuses on work limitations. A participant noted that, unfortunately, questions on work limitation have been eliminated from two basic population surveys, the American Community Survey (ACS) and the Current Population Survey (CPS) monthly questions, possibly because of a limitation of space, but also because of distrust of that methodology.

Barbara Altman explained that ACS and CPS have not eliminated questions on work, but only on work limitations. The reasoning was that to get work limitations, one can cross-tabulate people with functioning limitations with whether or not they work. ACS questions are going to be used in CPS; they are also going to be used in the National Crime Victimization Survey for the Department of Justice and in the National Health Interview Survey of the National Center for Health Statistics.

She agreed that the age focus needs to be expanded if any kind of preventive intervention is considered. She also identified some other areas that should be considered for addition or expansion. From a population survey perspective, the function and assessment aspects of the questions need to be expanded in order to expand one's understanding of the population that is at risk. Organized activities need to be added to population surveys on a regular basis; organized activities that are important to people's participation in multiple ways (such as transportation) give a clue about the kinds of barriers that a person might face. Finally, there is need for longitudinal data on populations.

Richard Suzman pointed out that cross-sectional population surveys are different from the collection of longitudinal data and one would not necessarily use the same questions. Also, different surveys have different purposes, and one should not impose a cross-sectional official survey framework on a research question.

Purpose of a Survey

Jay Magaziner observed that we need to ask why we are asking these measurement questions. What is it that we really want to answer? If the focus is issues that relate to underlying diseases and prevention of disability, then we need to develop questions and studies around that. If the purpose is having an item that can be used nationally to represent disability for policy purposes, that may be a different kind of question and a different set of items to make the responses more interpretable and useful. So, when talking about disability, it is important to think about the purpose of the inquiry, whether it is an individual research study or a global population survey.

Measures of Participation and the Environment

Alan Jette (Boston University) noted that if a key issue is the extent to which loss of function and diseases of late life affect the lives of individuals in society, then it is important to attempt to measure participation, at least in some dimensions. This is particularly important if one wants to assess the effectiveness of interventions for older populations at the societal level. The ability of older people to maintain engagement in important social roles in their lives is an extremely critical element, and Jette said that he sees it as part of the disablement process. There is a role for it in population surveys, and it can be measured. There is some evidence now that with techniques such as the item response theory and computer-adaptive testing (CAT), one can begin to get at some elements of important social roles in late life in a way that might be feasible to build into population surveys.

A participant commented that a barrier to obtaining better measures of participation and the environment in population surveys has been the burden of including such a large number of questions in a survey questionnaire. Given these new methodologies, is there now an opportunity to make these measurements more relevant, given the new psychometric techniques? The work could really benefit from short forms or CAT.

Jette responded that it is important to have improved measures of the environment, about which even less is known than in the measurement of participation. Measurements of the environment, if incorporated into population surveys, will help provide a better handle on the question: If the environment is changed, will it have a beneficial impact on the lives of older individuals? However, measures of the environment are not yet ready for incorporation in population surveys, but that is an important area for research.

Vicki Freedman noted that the environment does not mean only physical challenge, that is, it is not just the steps of the grocery store, if there are any, or how wide the aisles are, because technology has changed the meaning of the environment. For example, people can pick up the telephone to order prescriptions and have them delivered by the drugstore, which has changed the whole notion of environment.

It was pointed out that it is important to conceptualize the environment, not just as the physical environment, but also as the social environment and the public environment.

CONCLUDING COMMENTS

The workshop brought together a large group of participants from many different disciplines who presented and discussed the various approaches to measurement of disability, some tested and some still being tested. In this final session participants identified some of the major research areas and priorities that are emerging that they would like to see advanced by NIA and other organizations in terms of research and development, given the goal of improving the measurement of late-life disability in population surveys beyond ADLs and IADLs.

References

Banks, J., Breeze, E., Lessor, C., and Nazroo, J. (2006).(Eds.) *Retirement, Health and Relationships of the Older Population in England: The 2004 English Longitudinal Study of Ageing (Wave 2)*. London, England: Institute for Fiscal Stidies. Available: http://www.ifs.org.UK/elsa/report_wave2.php [accessed May 2009].

Barberger-Gateau, P., Fabrigoule, C., Rouch, I., Letenneur, L., and Dartigues, J.F. (1999). Neuropsychological correlates of self-reported performance in instrumental activities of daily living and prediction of dementia. *Journal of Gerontology: Pycholological Sciences* 54B(5):P293–P303.

Bassett, S.S., Magaziner, J., and Hebel, J.R. (1990). Reliability of proxy responses on mental health indices for aged community-dwelling women. *Psychology of Aging* 5(1): 127–132.

Brown, M., Dijkers, M., Gordon, W.A., Ashman, T., and Chartz, H. (2004). Participation objective, participation subjective: A measure of participation combining outsider and insider perspectives. *Journal of Head Trauma Rehabilitation* 19:459–481.

Brunner, E., Shipley, M., Spencer, V., Kivimaki, M., Chandola, T., Gimeno, D., Singh-Manoux, A., Guralnik, J., and Marmot, M. (2009). Social inequality in walking speed in early old age in the Whitehall II Study. *Journal of Gerontology: Medical Sciences* 64A(10): M1082-M1089.

Cella, D., Yount, S., Rothrock, N., Gershon, R., Cook, K., Reeve, B., Ader, D., Fries, J., Bruce, B., and Rose, M. (2007). The Patient-Reported Outcomes Measurement Information System (PROMIS): Progress of an NIH roadmap cooperative group during its first two years. *Medical Care* 45(5 Suppl 1):S3–S11.

Center for Data Collection and Research (CentERdata). (2003). *Data from the LISS Panel.* Available: http://www.centerdata.nl/en/TopMenu/Databank/LISS_panel_data/ [accessed April 2009].

Cherry Engineering Support Services, Inc. (2003). *Federal Statutory Definitions of Disability.* Prepared for the Interagency Committee on Disability Research. McLean, VA: Cherry Engineering Support Services, Inc. Available: http://www.icdr.us/documents/definitions.htm [accessed May 2009].

Cocchiarella, L., and Andersson, G. (2001). (Eds.). *Guides to the Evaluation of Permanent Impairment* (5th ed.). Chicago, IL: American Medical Association.

Cook, K.F., O'Malley, K.J., and Roddey, T.S. (2005). Dynamic assessment of health outcomes: Time to let the CAT out of the bag? *Health Services Research* 40(5 Pt 2):1694–1711.

Cornoni-Huntley, J.C., Foley, D.J., and Guralnik, J.M. (1991). Co-morbidity analysis: A strategy for understanding mortality, disability and use of health care facilities of older people. *International Journal of Epidemiology* 20:S8–S17.

Coster, W.J., Haley, S.M., Ludlow, L.H., Andres, P.L., and Ni, P.S. (2004). Development of an applied cognition scale to measure rehabilitation outcomes. *Archives of Physical Medicine and Rehabilitation* 85(12):2030–2035.

Couper, M., Kapteyn, A., Schonlau, M., and Winter, J. (2007). Noncoverage and nonresponse in an Internet survey. *Social Science Research* 36(1):131–148.

Crimmins, E.M., Saito, Y., and Reynolds, S. (1997). Further evidence on recent trends in the prevalence and incidence of disability among older Americans from two sources: The LSOA and the NHIS. *Journal of Gerontology: Social Sciences* 52B(2):S59–S71.

De Ayala, R.J. (1993). Methods, plainly speaking: An introduction to polytomous item response theory models. *Measurement and Evaluation in Counseling and Development* 25:172–189.

Di Carlo, A., Baldereschi, M., Amaducci, L., Maggi, S., Grigoletto, F., Scarlato, G., and Inzitari, D. (2000). Cognitive impairment without dementia in older people: Prevalence, vascular risk factors, impact on disability: The Italian Longitudinal Study on Aging. *Journal of the American Geriatrics Society* 48(7):775–782.

Donovan, N.J., Kendall, D.L., Heaton, S.C., Kwon, S., Velozo, C.A., and Duncan, P.W. (2008). Conceptualizing functional cognition in stroke. *Neurorehabilitation and Neural Repair* 22(2):122–135.

Ferrucci, L., Guralnik, J.M., Simonsick, E., Salive, M.E., Corti, C., and Langlois, J. (1996). Progressive versus catastrophic disability: A longitudinal view of the disablement process. *Journal of Gerontology: Medical Sciences* 51A(3):M123–M130.

Freedman, V. (2006). Late-life disability trends: An overview of current evidence. In Institute of Medicine, *Workshop on Disability in America: A New Look*. M. Field, A.M. Jette, and L. Martin (Eds.). Committee on Disability in America. Washington, DC: The National Academies Press.

Freedman, V.A., Martin, L.G., and Schoeni, R.F. (2002). Recent trends in disability and functioning among older adults in the United States: A systematic review. *Journal of the American Medical Association* 288:3137–3146.

Freedman, V.A., Crimmins, E., Schoeni, R.F., Spillman, B.C., Aykan, H., Kramarow, E., Land, K., Lubitz, J., Manton, K., Martin, L.G., Shinberg, D., and Waidmann, T. (2004). Resolving inconsistencies in trends in old-age disability: Report from a technical working group. *Demography* 41(3):417–441.

Freedman, V.A., Waidmann, T., and Spillman, B. (2005). *Opportunities to Improve Survey Measures of Late-Life Disability: Part II—Workshop Summary.* Report submitted to the Assistant Secretary for Planning and Evaluation, U.S. Department of Health and Human Services, Washington, DC.

Fried, L.P., and Herdman, S.J. (1992). Geriatric assessment: Multifactorial problems. *Physical Therapy Practice* 1:88–93.

Fried, L.P., Borhani, N.O., Enright, P., Furberg, C.D., Gardin, J.M., Kronmal, R.A., Kuller, L.H., Manolio, T.A., Mittelmark, M.B., Newman, A., O'Leary, D.H., Psaty, B., Rautaharju, P., Tracy, R.P., and Weiler, P.G. (1991a). The Cardiovascular Health Study: Design and rationale. *Annals of Epidemiology* 1:263–276.

Fried, L.P., Herdman, S.J., Kuhn, K., Rubin, G., and Turano, K. (1991b). Preclinical disability: Hypotheses about the bottom of the iceberg. *Journal of Aging and Health* 3:285–300.

Fried, L.P., Bandeen-Roche, K., Chaves, P.H., and Johnson, B.A. (2000). Preclinical mobility disability predicts incident mobility disability in older women. *Journal of Gerontology: Medical Sciences* 55A(1):M43–M52.

Gill, T.M., and Gahbauer, E. (2008). Evaluating disability over discrete periods of time. *Journal of Gerontology: Medical Sciences* 63A(6):M588–M594.

Gill, T.M., Robinson, J.T., and Tinetti, M.E. (1998). Difficulty and dependence: Two components of the disability continuum among community-living older persons. *Annals of Internal Medicine* 128:96–101.

Gill, T.M., Hardy, S., and Williams, C. (2002). Underestimation of disability in community-living older persons. *Journal of the American Geriatrics Society* 50:1492–1497.

Gill, T.M., VanNess, P., and Gahbauer, E. (in press). Factors associated with accurate recall of prior disability in older persons. *Journal of the American Geriatrics Society.* Early view available: http://www3.interscience.wiley.com/cgi-bin/fulltext/122592466/PDFSTART [accessed September 2009].

Hambleton, R. (2005). Applications of item response rheory to improve health outcomes assessment: Developing item banks, linking instruments, and computer adaptive testing. In J. Lipscomb, C. Gotay, and C. Snyder (Eds.), *Outcomes Assessment in Cancer* (pp. 445–464). Cambridge, England: Cambridge University Press.

Harris, T., Kovar, M.G., Suzman, R., Kleinman, J.C., and Feldman, J.J. (1989). Longitudinal study of physical ability in the oldest-old. *American Journal of Public Health* 79(6):698–702.

He, W., Sengupta, M., Velcoff, V.A., and DeBarros, K.A. (2005). *65+ in the United States: 2005.* (U.S. Census Bureau, Current Population Reports, P23–209.) Washington, DC: U.S. Department of Commerce.

Iburg, K., Salomon, J., Tandon, A., and Murray, C. (2001). *Cross-Population Comparability of Self-Reported and Physician-Assisted Mobility Levels: Evidence from the Third National Health and Nutrition Examination Survey.* (Global Programme on Evidence for Health Policy Discussion Paper No. 14.) Geneva, Switzerland: World Health Organization.

Iezzoni, L., and Freedman, V. (2008). Turning the disability tide: The importance of definitions. *Journal of the American Medical Association* 299:332–334.

Institute of Medicine. (1991). *Disability in America: Toward a National Agenda for Prevention.* A.M. Pope and A.R.Tarlov (Eds.). Washington, DC: National Academy Press.

Institute of Medicine. (1997). *Enabling America: Assessing the Role of Rehabilitation Science and Engineering.* E.N. Brandt, Jr., and A.M. Pope (Eds.). Washington, DC: National Academy Press.

Institute of Medicine. (2007). *The Future of Disability in America.* M.J. Field and A.M. Jette (Eds.). Washington, DC: The National Academies Press.

Ishizaki, T., Yoshida, H., Suzuki, T., Wantanabe, S., Niino, N., Ihara, K., Kim, H., Fujiwara, Y., Shinkai, S., and Imanaka, Y. (2006). Effects of cognitive function on functional decline among community-dwelling nondisabled older Japanese. *Archives of Gerontology and Geriatrics* 42(1):47–58.

Jette, A.M., and Badley, E. (2002). Conceptual issues in the measurement of work disability. In Institute of Medicine and National Research Council, *The Dynamics of Disability: Measuring and Monitoring Disability for Social Security Programs* (pp. 53–64). G.S. Wunderlich, D.P. Rice, and N.L. Amado (Eds.). Committee to Review the Social Security Administration's Disability Decision Process Research. Washington, DC: National Academy Press.

Kahneman, D., Krueger, A.B., Schkade, D., Schwarz, N., and Stone, A.A. (2004). A survey method for characterizing daily life experience: The Day Reconstruction Method (DRM). *Science* 306(5702):1776–1780.

Kapteyn, A., Smith, J.P., and van Soest, A. (2007). Vignettes and self-reports of work disability in the United States and the Netherlands. *American Economic Review* 97(1):461–473.

Katz, S., and Akpom, C. (1976). A measure of primary sociobiological function. *International Journal of Health Services* 6:493–508.

King, G., Murray, C.J.L., Salomon, J.A., and Tandon, A. (2004). Enhancing the validity of cross-cultural comparability of measurement in survey research. *American Political Science Review* 98(1):91–207.

Krueger, A., and Stone, A. (2008). Assessment of pain: A community-based diary survey in the USA. *Lancet* (371):1519–1525.

Kumamoto, T., Sannomiya, K., Ueyama, H., Aoki, K., Nakashima, T., Nakamura, R., and Tsuda, T. (2000). Neurological abnormalities in cognitively impaired but not demented elderly. *Acta Neurologica Scandinavia* 102(5):292–298.

Lakdawalla, D.N., Bhattacharya, J., and Goldman, D. (2004). Are the young becoming disabled? *Health Affairs* 23(1):168–176.

Lan, T-Y., Melzer, D., Tom, B., and Guralnik, J. (2002). Performance tests and disability: Developing an objective index of mobility-related limitation in older populations. *Journal of Gerontology, Medical Sciences* 57A(5): M294–M301.

Lawton, M.P., and Brody, E. (1969). Assessment of older people: Self-maintaining and instrumental activities of daily living. *Gerontologist* 9:179–186.

Magaziner, J., Bassett, S.S., and Gruber-Baldini, A. (1996). Use of proxies to measure health and functional status in epidemiologic studies of community-dwelling women aged 65 years and older. *American Journal of Epidemiology* 143(3):283–292.

Manton, K.G., and Gu, X. (2001). Changes in the prevalence of chronic disability in the United States black and nonblack population above age 65 from 1982 to 1999. *Proceedings of the National Academy of Sciences* 98(11):6354–6359.

Manton, K.G., Corder, L., and Stallard, E. (1997). Chronic disability trends in elderly United States populations: 1982–1994. *Proceedings of the National Academy of Sciences* 94: 2593–2598.

Melzer, D., Lan, T-Y., Tom, B., Deeg, D., and Guralnik, J. (2004). Variation in thresholds for reporting mobility disability between national population subgroups and studies. *Journal of Gerontology: Medical Sciences* 59A(12):M1295–M1303.

Miller, D.K., Wolinsky, F.D., Malmstrom, T.K., Andresen, E.M., and Miller, J.P. (2005). Inner city, middle-aged African Americans have excess frank and subclinical disability. *Journal of Gerontology: Medical Sciences* 60(2):M207–M212.

Nagi, S. (1965). Some conceptual issues in disability and rehabilitation. In M. Sussman (Ed.), *Sociology and Rehabilitation* (pp. 100–113). Washington, DC: American Sociological Association.

Nagi. S. (1976). An epidemiology of disability among adults in the United States. *Milbank Memorial Fund Quarterly* 54:439–467.

Nagi, S. (1991). Disability concepts revisited: Implications for prevention. In Institute of Medicine, *Disability in America: Toward a National Agenda for Prevention* (pp. 309–327). A.M. Pope and A.R. Tarlov (Eds.). Washington, DC: National Academy Press.

National Center for Health Statistics. (2009). *Health, United States, 2008. Chartbook with Special Feature on the Health of Young Adults.* Hyattsville, MD: National Center for Health Statistics.

National Institute on Aging and National Bureau of Economic Research. (2001). *Disability Decline: What We Know and What We'd Like to Know. Workshop Summary.* Available: http://www.nia.nih.gov/ResearchInformation/ExtramuralPrograms/BehavioralAndSocialResearch/ConferencesAndWorkshops.htm [accessed February 2009].

Purser, J.L., Fillenbaum, G.G., Pieper, C.F., and Wallace, R.B. (2005). Mild cognitive impairment and 10-year trajectories of disability in the Iowa Established Populations for Epidemiologic Studies of the Elderly cohort. *Journal of the American Geriatrics Society* 53(11):1966–1972.

Raji, M.A., Kuo, Y.F., Snih, S.A., Markides, K.S., Peek, M.K., and Ottenbacher, K.J. (2005). Cognitive status, muscle strength, and subsequent disability in older Mexican Americans. *Journal of the American Geriatrics Society* 53(9):1462–1468.

Reuben, D.B., Laliberte, L., Hiris, J., and Mor, V. (1990). A hierarchical exercise scale to measure function at the advanced activities of daily living level. *Journal of the American Geriatrics Society* 38:855–861.

Rosow, I., and Breslau, N. (1966). A Guttman health scale for the aged. *Journal of Gerontology* 21:556–559.

Schoeni, R.F., Freedman, V.F., and Wallace, R. (2001). Persistent, consistent, widespread, and robust? Another look at recent trends in old age disability. *Journal of Gerentology: Social Sciences* 56B(4):S206–S218.

Schupf, N., Tang, M-X., Albert, S.M., Costa, R., Andrews, H., Lee, J.H., and Mayeux, R. (2005). Decline in cognitive and functional skills increases mortality risk in nondemented elderly. *Neurology* 65(8):1218–1226.

Shumway-Cook, A., Ciol, M.A., Yorkston, K.M., Hoffman, J.M., and Chan, L. (2005). Mobility limitations in the Medicare population: Prevalence and sociodemographic and clinical correlates. *Journal of the American Geriatrics Society* 53(7):1217–1221.

Social Security Administration. (2003). *Disability Evaluation under Social Security.* (SSA Publication No. 64–039.) Washington, DC: Social Security Administration, Office of Disability Programs.

Soldo, B.J., Mitchell, O.S., Tfaily, R., and McCabe, J.F. (2006). *Cross-Cohort Differences in Health on the Verge of Retirement.* (NBER working paper no. 12762.) Cambridge, MA: National Bureau of Economic Research. Available: http://www.nber.org/papers/w12762 [accessed September 2009].

Steen, G., Sonn, U., Hanson, A.B., and Steen, B. (2001). Cognitive function and functional ability: A cross-sectional and longitudinal study at ages 85 and 95 in a nondemented population. *Aging (Milano)* 13(2):68–77.

U.S. Department of Health and Human Services. (1989). *Measuring the Activities of Daily Living Among the Elderly: A Guide to National Surveys.* Report prepared by J.M. Weiner and R.J.M. Hanley, Brookings Institute, for the Federal Interagency Forum on Aging-Related Statistics.

Verbrugge, L., and Jette, A.M. (1994). The disablement process. *Social Science and Medicine* 38(1):1–14.

Waidmann, T., and Freedman, V. (2007). *The Daily Activities of the Community-Dwelling Elderly: Evidence from the American Time-Use Survey.* Report to the U.S. Department of Health and Human Services, Washington, DC. Available: http://aspe.hhs.gov/daltcp/reports/2006/cdelderly.pdf [accessed May 2009].

Waidmann, T., and Liu, K. (2000). Disability trends among the elderly and implications for the future. *Journal of Gerentology: Social Sciences* 55B:S298–S307.

Ware, J.E., Jr., Bjorner, J.B., and Kosinski, M. (2000). Practical implications of item response theory and computerized adaptive testing: A brief summary of ongoing studies of widely used headache impact scales. *Medical Care* 38(9 Suppl):1173.

Weiss, C.O., Hoenig, H.M., and Fried, L. (2007). Compensatory strategies used by older adults facing mobility disability. *Archives of Physical Medicine and Rehabilitation* 88(9): 1217–1220.

Whiteneck, G.G., Charlfue, S.W., Gerhart, K.A., Overholser, J.D., and Richardson, G.N. (1992). Quantifying handicap: A new measure of long-term rehabilitation outcomes. *Archives of Physical Medicine and Rehabilitation* 73:519–526.

Wiener, J.M., Hanley, J.R., Clark, R., and Van Nostrand, J. (1990). Measuring the activities of daily living: Comparisons across national surveys. *Journal of Gerontology: Social Sciences* 45(6):S229–S237.

Wolinsky, F.D., Miller, D.K., Andresen, E.M., Malmstrom, T.K., Miller, J.P., and Miller, T.R. (2007). Effect of subclinical status in functional limitation and disability on adverse health outcomes 3 years later. *Journal of Gerontology: Medical Science* 62(1):101–106.

World Health Organization. (2001). *International Classification of Functioning, Disability and Health*. Geneva, Switzerland: World Health Organization.

World Health Organization. (2002). *Towards a Common Language for Functioning, Disability, and Health: International Classification of Functioning, Disability and Health*. Geneva, Switzerland: World Health Organization.

Wu, X.G., Tang, Z., Fang, X.H., Liu, H.J., Diao, L.J., and Xiang, M.J. (2004). Evaluation of predictive effect of some health-related indices on deaths among ageing residents through 8-years' follow-up study in Beijing. *Zhonghua Liu Xing Bing Xue Za Zhi* 25(4):325–328.

Appendix A

Population Survey Measures of Functioning: Strengths and Weaknesses*

Barbara M. Altman

INTRODUCTION

Thomas (2002) has pointed out that "disability" is a common term whose meaning is seen as beguilingly obvious to the layperson—not being able to do something. However, the legislative use of the term in policy and the scientific use of the term in medical, rehabilitation, or welfare settings to define the relevant population is fraught with shades of meaning that vary according to the purposes of those activities. Meaning is also influenced by the way people conceptualize and measure within those purposes. These multiple, as well as different, approaches to the understanding of the terms of discussion obscure theoretical definitions and confuse communication when one tries to examine the big picture.

Since scientific disciplines tend to work within the bounds of their own literatures, terminology, and models, these variations in definition, measurement, and meaning persist. The history of the involvement of many disciplines in the development of policy and programs to address disability issues—medicine, education, social work, psychology, sociology, vocational counseling, occupational and physical therapy, and others—sets the stage for attempts at conceptual distinctions to delineate measures of the disability process (Nagi, 1991). Importantly, even when definitional components are generalized and accepted across fields (such as with the International

*The author gratefully acknowledges the meaningful input on conceptualization, organization, and editing provided by Julie Weeks of the National Center for Health Statistics of the Centers for Disease Control and Prevention.

Classification of Functioning, Disability and Health [ICF]), measurement of those components in terms of disciplinary practice reintroduces different shades of meaning and alters the conceptual definition in subtle but specific ways.

The purpose of this paper is threefold. First, I explore the disability conceptualization/measurement continuum in the area of population surveys, to locate it within the general theory of disability and to compare it across specialty areas of application. Second, I define and locate the source and types of measurement for the variety of theoretical concepts associated with the disability process. Finally, I reintroduce the important contribution of the social and environmental context not only to the conceptualization of disability, but also to its measurement.

THE MEDICAL CONTEXT

Among other things, the lay "beguiling" meaning of disability generally locates the source of the problem within individuals, based solely on their physical or mental impairment or imperfection. This personal attribution comes in part from the medical orientation to chronic conditions and the physical or mental impairment consequences of those conditions which focus both diagnosis and treatment on individuals. Although diagnosis and treatment *are* appropriately directed toward the chronic condition or impairment a person exhibits (consideration of the individual as the host of a problem), locating the larger phenomenon of "disability" within the individual in the same way is inappropriate and misrepresents the situation. There are actually at least three major component areas that contribute to creating a disabling outcome, and none of them is located *within* the person.

The medical model, which has influenced so many of the nation's disability-related programs, is not so much a separate theoretical model but the primary context in which the available theoretical models have been interpreted through research and applied in programmatic activity. The emphasis is placed on the physical or mental impairment within the person, assuming it is the sole factor that somehow constrains activity and access. The medical model begins in a medical setting, with the emphasis on diagnosis, treatment, and rehabilitation of people as they move from active pathology to impairment. Given the medical orientation, the medical model also is a form of data collection, such that the source of data is a physical examination or observed performance.

Following the initial stages of the disability process that grow out of the relationship with medical organizations, disability-related programs, many of which were structured in an earlier time, continue the focus on the individual, ignoring the culpability of society and the environment. So, for example, the organization of the Social Security program was focused on replacing the person's monetary support base and not on retraining them for a different kind of job or providing a legislative agenda to prevent discrimination in the job market. With the emphasis on replacing income, stringent eligibility criteria, which required medical documentation, were necessary because of the potential long-term costs of the program. Vocational rehabilitation continued this focus on the person but broadened it to include training, medical care, and therapy—all directed at the person—rather than seeking to remove structural and social barriers by rehabilitating the workplace as well.

As early as the late 1980s, individuals and groups were pointing to problems with emphasizing the individual in the disability equation (Caplan, 1988). It was a very hard sell; however, *Disability in America* (Institute of Medicine, 1991) did give space to the idea. Nonetheless, the emphasis in that groundbreaking work was still the person, not only the conditions causing disability, but also the problem of secondary conditions that could make a disability worse. Because the mandate of that volume was to explore the *prevention* of potentially disabling conditions from developing into disability at every stage of the disability process, the emphasis was heavily weighted toward what could be done for the person in the medical context, from the prevention of secondary conditions to the use of rehabilitative and vocational services.

The contribution of Verbrugge and Jette (1994) in their elaboration of the disability models in use at the time (Nagi, 1965; World Health Organization, 1980) was the beginning of bringing the understanding of the disability process beyond the medical context. Moving the theoretical models beyond the medical context laid the groundwork for also moving the data collection about the process into a social survey context. The expanded 1996 Medical Expenditure Panel Survey and the 1994–1995 National Health Interview Survey on Disability (NHIS-D) solidified that position. They also added nationally representative data on disability, which included broader sets of questions and the whole population, to what had previously been specialty surveys on particular populations, such as Social Security recipients, children, or the aging population.

CURRENT CONCEPTUALIZATIONS OF DISABILITY

Nagi Model

Several models allow us to examine the elements of disability and explore where they are located and how they relate to one another. The original model that gained widespread acceptance was authored by Nagi and included the conceptual elements of pathology, impairment, functional limitation, and disability (Nagi, 1965). Although Nagi never presented his theoretical model in a symbolic format, others have, implying a unidimensional progression from pathology to an outcome called "disability." However, a close and careful reading of the original Nagi theoretical exposition (Nagi, 1965) and later works (Nagi, 1969, 1991) can be interpreted somewhat differently, symbolically as well as conceptually. Moreover, key constructs presented by Nagi—role interactions and behaviors—are omitted altogether from most secondary interpretations. I have elaborated the symbolic representation accounting for the role interaction situation Nagi originally postulated and emphasizing the individual's behaviors that follow. It is this pattern of behavior that Nagi defined as disability (Altman, 2001b). These conceptual subtleties, which are frequently glossed over, are discussed here.

There are several important points to take from the Nagi model that are relevant to the definition and measurement of disability in survey situations (see Figure A-1). First, disability is not the equivalent of an individual's conditions or impairments, be they pathologies, residual losses, or abnormalities caused by injury, disease, or congenital accident. Conditions and or impairments are attributes of the individual that eventually affect the nature and degree of disability through their effects on the person's overall

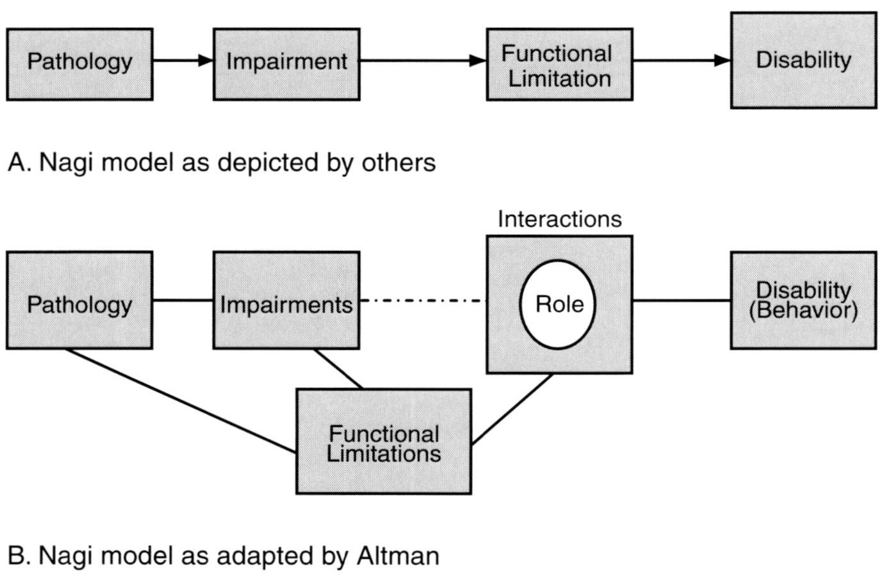

FIGURE A-1 Two versions of Nagi model.
SOURCE: Altman (2001b).

functioning. Rather, disability refers to *social* outcomes. It is related to performing socially defined roles,[1] from self-care to employment, and is shaped by the expectations, reactions, and definitions of people around the "disabled" individual, as well as by the impact of the physical environment. It reveals itself most completely in the *behaviors* exhibited by an individual in social situations (Nagi, 1965).

Although Nagi changed this definition of disability somewhat in later work (Nagi, 1991) to reflect a more social functioning orientation, the broader orientation of his earlier work is used in this paper as well as the later one because of its importance in measurement. Although the differences between "functioning" in a role and "behavior" in a role may seem quite subtle, they are important differences in understanding the full effects of the environmental contribution to disability. Questions based on a functioning perspective still locate the problem primarily in the person's ability to do the activity within the given context, such as being limited in the kind or amount of work or unable to work. Questions about behavior can more accurately reflect the adjustments that are made both by the person and in the context and also reflect the person's *choice* in the situation. Functioning reflects assessment of ability or capacity in the general situation, whereas behavior is the manifestation of the actual interaction that creates disability. While functioning in basic actions, such as walking or seeing, is an important component of the equation at the person level, functioning in the area of a more complex activity, such as working, focuses on the outcome of the person–environment interaction. This latter functioning tells us what the person can do in such a situation or the difficulty they have doing it, but it tells us nothing about how their behavior has been modified to accommodate the barriers or how the barriers have been modified to accommodate the functional limitation.

The strength of the Nagi model lies in its insights about the whole process, which still have relevance after more than 40 years. Two areas are particularly important. First, the recognition of the difference between impairment, the damaged organ, cell, or body part, and functioning, actions of the whole person, which focuses the measurement on the person's capacity rather than the type of condition extant within the person. This differen-

[1]This is an example of how disciplinary language and orientations impose subtlely but importantly on a conversation of this type. Nagi is a sociologist, and his work reflects a sociological orientation to culture and social roles. The general term social role (or simply role) is used to refer to both a position and its associated behavioral expectations. The role of a person with a functional limitation as understood through the behavioral expectations of the group of which they are a part is a key factor in creating disability.

tiation identifies functioning as the fulcrum in the disability process.[2] The other area of strength is acknowledgment of the important role that social interaction with friends, family, and the community contributes to defining the situation. Although the physical environment is not mentioned in the original discussion of the model, it is certainly addressed in the study of disability and rehabilitation, the source of the "Nagi questions," which were written a few years later (Nagi, 1969). The weakness seems to be the lack of further conceptual elaboration of the behavioral component sufficient to provide clues to its measurement.

Institute of Medicine Models

Two other commonly referenced models are the Institute of Medicine (IOM) model from *Enabling America* (Institute of Medicine, 1997) and its predecessor from *Disability in America* (Institute of Medicine, 1991). Both are products of the IOM committee process. Both are based on the Nagi model and the later version is very much influenced by adaptations of the Nagi model by Verbrugge and Jette (1994).

The 1997 IOM model depicted in Figures A-2 and A-3 follows the Nagi model concepts closely in the areas of pathology, impairment, and functional limitation and collapses some of the Verbrugge and Jette elements of intra- and extra-individual factors into the quality-of-life components (Verbrugge and Jette, 1994). Disability is addressed separately (see Figure A-3) from the overall model to emphasize the interactive nature of its creation and, one assumes, to focus attention on the complexity of a process that is easily ignored. What is indicated clearly in this part of the model is that disability is not inherent in the person, but rather a result of the interaction of the individual and the environment. In other sections of the IOM text, reference is made to adjustments to the physical environment or rehabilitation of the individual, both of which can contribute to modifying the level of disability a person experiences.

The strengths of the IOM model include its reinforcement of the Nagi separation of impairment and functioning and the provision of a much clearer presentation of the person–environment interaction that leads to disability. The latter is particularly meaningful for measurement purposes because it broadens the measurement potential beyond individual functioning and an independence–dependence dichotomy—represented by activities of daily living (ADLs) and instrumental activities of daily living (IADLs)—to the person–environment context. The IOM discussion acknowledges the

[2]From the author's perspective it also takes the process out of the medical context so important for the individual's health but separate from the social construction of disability that results through interpersonal, organizational, and social activities (Bogdan and Biklen, 1977).

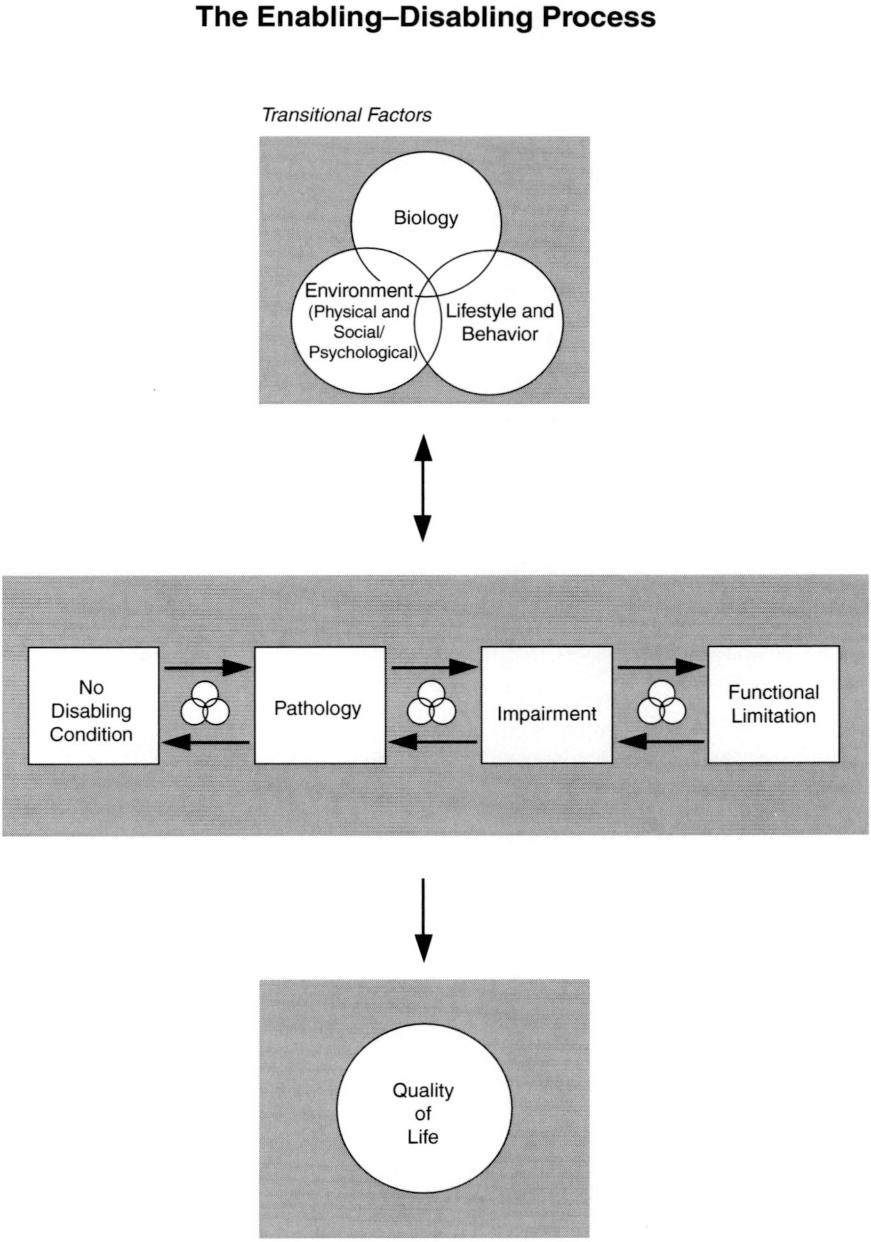

FIGURE A-2 Second Institute of Medicine model.
SOURCE: Institute of Medicine (1997), adapted from the first model (Institute of Medicine, 1991).

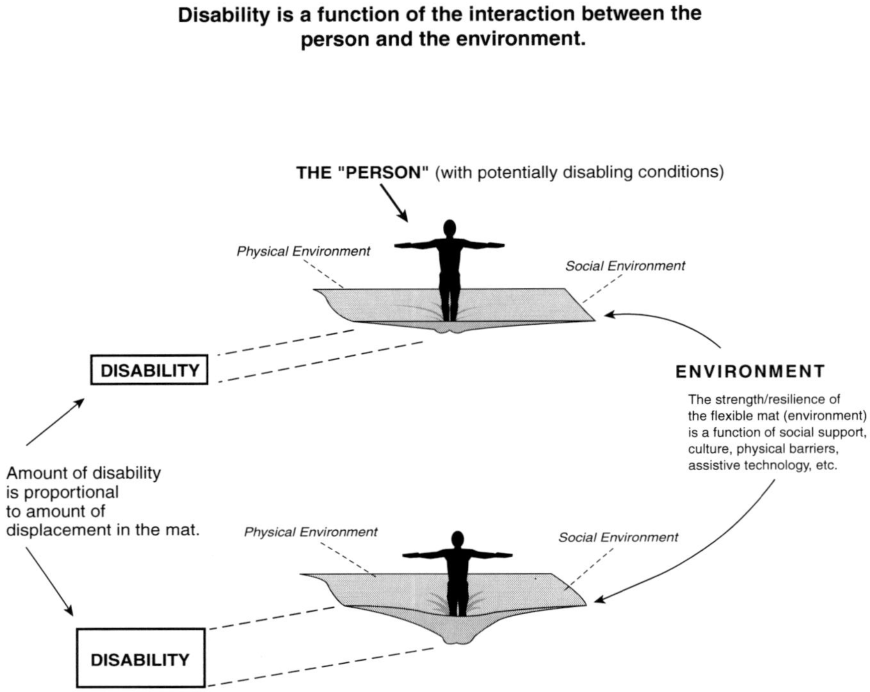

FIGURE A-3 Complementary component of Institute of Medicine model: Conceptual overview of the person–environment interaction and how that interaction creates the disability.
SOURCE: Institute of Medicine (1997).

multidisciplinary character of the study of disability by very efficiently pointing to cultural, psychological, economic, and political questions that need to be addressed (Institute of Medicine, 1997, p. 168). However, although the model moved forward an understanding of the process, particularly the contribution of the external environment, it lacked insights into actual mechanisms associated with this environment–person interaction. This leaves a gap in the measurement process for environmental elements without a clear direction for the conceptual operationalization process. As with the ICF model that follows, the importance of environment is clear, but where to start the measurement process is vague.

International Classification of Functioning, Disability and Health Model

ICF (World Health Organization, 2001a) is an extensive revision of the International Classification of Impairments, Disabilities, and Handicaps

(World Health Organization, 1980), both products of the World Health Organization. This final model of the disability process currently in use (shown in Figure A-4), like the IOM model, does not include the state of "disability" within the outlines of the conceptual relationships. It also does not include the concept of functioning limitation so prominent in the other two models.

Disability in this model is an umbrella term for the *negative* aspects of the interaction between an individual and that individual's contextual factors, while functioning is the umbrella term used to denote the *positive* aspects of the interaction between an individual in his or her context. The conceptual elements in this model reflect, for the most part, the elements of body structure and function (identifying the impairment as well as the body system or body part where it is located), activity (which represents a person's execution of a task or action—the individual perspective of functioning), and participation (which represents a person's involvement in a life situation, functioning at a complex level of activity). However, there are continuing problems differentiating the activity and participation components for use when applying the conceptual model in practice, as in the case of measurement (Chapireau, 2005; Whiteneck, 2006; Institute of Medicine, 2007). Environmental factors are noted, along with personal factors as modifiers of the whole set of concepts similarly to the use of biology, environment, lifestyle, and behaviors as risk factors in the IOM model.

The ICF model, which has received international support, also provides a taxonomy for classifying function, disability, and health with standard

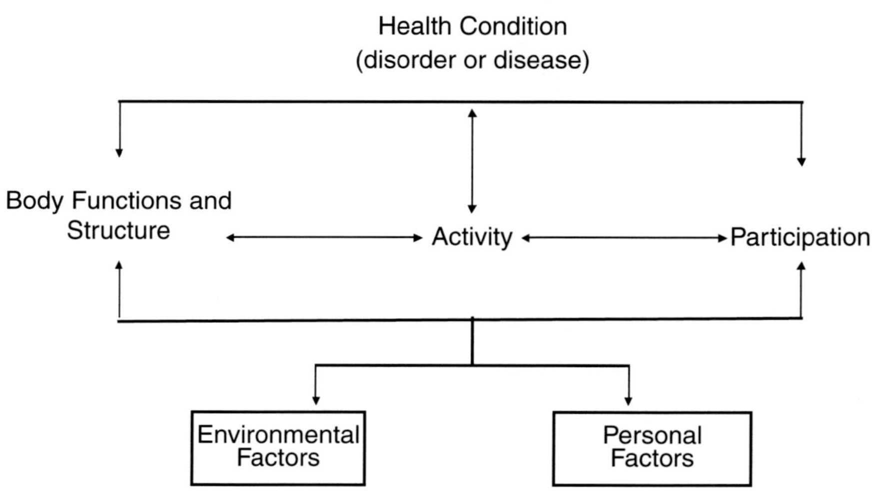

FIGURE A-4 ICF model.
SOURCE: World Health Organization (2001a).

concepts and terminology. It provides an extensive set of domains that represent the three primary areas of the model: body function and structure, activity and participation, and environmental factors. The domains are accompanied by a coding scheme that allows for classification of a person's characteristics at four different levels of specificity. In addition, there is also a coding scheme to denote qualifiers to represent the level of severity of the problem in the case of an impairment or the degree of barrier located in the environment. There are also optional qualifiers and a set of rules for their application. These domains and codes are posited as a measuring function, but it is misleading: although they may be useful in clinically oriented measurement associated with the body function and body structure sections, they are open to wide differences in operationalization in the activity and participation sections.

ICF moves the field forward through the standardization of some terminology and the specification of domains or areas that can be considered points of operationalization for measurement purposes. However, the use of disability and functioning as umbrella terms creates as much confusion as clarity when they are used to represent everything. The emphasis on the body function and structure compliments the health condition classification of ICD-10 and makes the ICF classification a promising tool for clinical settings. However, from a social survey perspective, this emphasis on body structure and function, combined with the lack of differentiation between activity and participation, confuses the development of measurement to represent the process. The environmental context is viewed so broadly as to provide little guidance for measurement purposes.

All three models are similar in that they identify some form of impairment, individual functioning, and societal participation in an environmental context as the key components involved in disability (some more clearly and others more broadly). The ICF model has difficulty providing a clear demarcation between activity and participation, but it provides a classification schema with an extensive set of domains to be considered. Although the Nagi and IOM concepts of functioning are directed more specifically at the whole person's basic actions than is the ICF, the Nagi concept of "disability" and the IOM concept of "disablement" are less specific and much more open to a wide range of possible interpretations from a measurement point of view. Environment in all three models is expressed in very broad brushstrokes, with emphasis in the Nagi model on the interaction with significant others or role sets;[3] in the IOM model on physical modifications

[3]Role set refers to the persons most commonly found to interact with an individual in specific role situations, so in the family situation the role set includes a spouse and children or parents, whereas in an employment situation a role set includes employers and colleagues or coworkers.

and vague definitions of social factors; and in the ICF model to all aspects of the extrinsic world that can contribute to the context in various settings. The last, though accurate, makes it nearly impossible to select a meaningful measure in a simple, concise way necessary in survey work.

FROM CONCEPT TO MEASUREMENT

Locating the Point of Measurement

The first concept considered here is the pathology or condition that is the causative health problem for creating whatever impairment may result. This element is located within the person's physical, mental, or emotional organs or organ systems; see Figure A-5. The pathology or condition is a health problem, an injury, or a congenital defect. It is the focus of diagnosis and treatment in the medical context. It takes place or is found in the person in one or more sites in or on the body. It is not a disability, although it can be a precursor to disability by causing an impairment that can in-

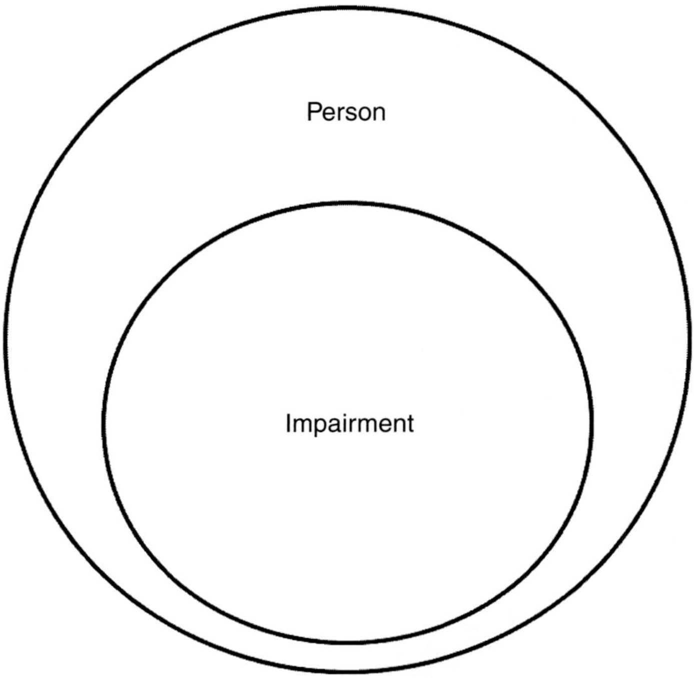

FIGURE A-5 Impairment, located within a person.

hibit functioning. So in the Nagi symbolic model (as adapted; see Altman, 2001b) and the IOM model, both the pathology and the impairment are located at the person level. In the ICF model, health or disease is presented as an intervening variable, and body function and structure represent the impairment and its location as the link with the International Classification of Disease. All three models identify the pathology, condition, and impairment in the person; thus measurement should be operationalized at the person level.

The second concept for consideration is limitation in functioning; see Figure A-6. In Nagi's interpretation these are "limitations which impairments set on the individual's ability to perform the tasks and obligations of the usual roles and normal daily activities" (Nagi, 1965, p. 102). The

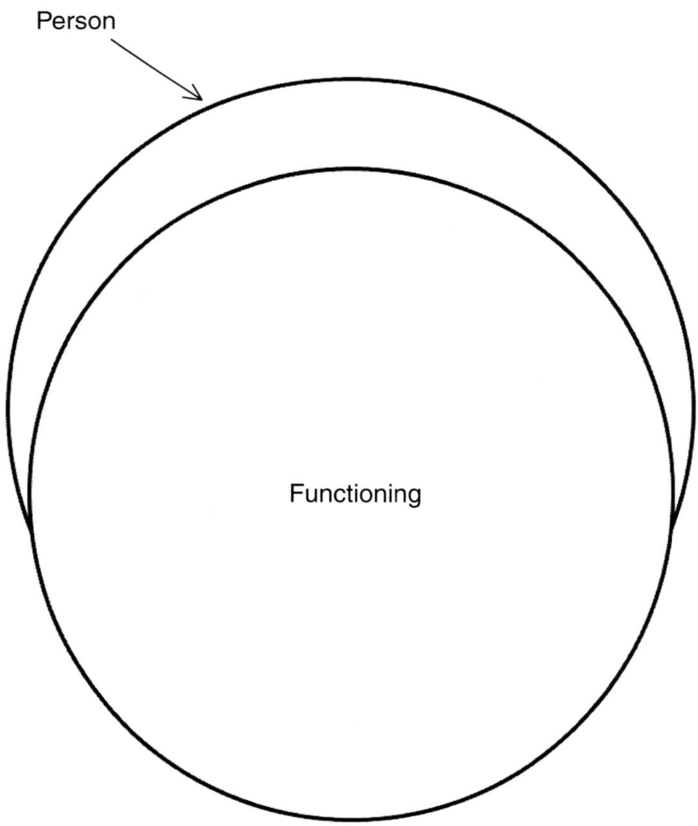

FIGURE A-6 Functioning, located at the level of a person.

emphasis is on the individual performing the basic actions necessary to accomplish everyday tasks, roles, and obligations. This refers to the ability of a person to perform the building blocks of action, such as walking, communicating, making decisions, and so forth. This type of functioning difficulty is not condition or impairment specific, but it may be common to several different types of conditions or impairments. For example, people with severe heart disease can have problems walking any distance, which is similar to people with orthopedic problems. In this instance, the conceptual focus is on the whole person and the functioning or accomplishment of the basic action, which requires coordination of physical, mental, and emotional capacities, and the choice or will to do it is concomitant with the person rather than one or more body parts. Mechanical aids to help with functioning are useful tools, but they reflect an environmental accommodation. In terms of measurement, functioning should be assessed without the use of aids in order to reflect the person's fundamental capacity.

The ICF model collapses actions and tasks into one conceptual component, activity. It does not separate the physical, cognitive, and emotional functioning from the tasks that require such individual functioning to be accomplished. The Nagi idea of individual functioning can be considered to be captured in the activity conceptualization of the ICF model, but the activity concept includes much more as well. In the ICF chapter on mobility, moving around is not only conceptualized as the basic action of *walking* but also as the organized activity of *using transportation* to move around, which requires a combination of different types of individual functioning ability other than just the basic ability to walk. Although this is an adequate approach to conceptualizing a hierarchy for a classification scheme, it confuses the hierarchy of a measurement process. Walking can be understood as a basic building block of action at the person level, but accomplishing a task, such as using transportation, in addition to combining the capacity to walk with other actions (such as cognitive functioning) to locate transportation, also requires making a choice as to what kind of transportation, involves seeing or hearing in order to use a schedule and so forth. Thus, using a transportation system not only represents using multiple areas of functioning, but also is influenced by a cultural component that creates the types and availability of transportation as well as the expectations for how those forms of transportation are to be used. It represents a much more complex activity; and thus involves increased issues associated with measuring (see Figure A-6).

The complement to this functioning capacity, which Nagi (1965) addresses but is omitted from newer models of disability, is the cultural component, the agreed-on nature, requirements, and expectations of the roles and activities a person is attempting. So, for example, bathing in a developing country may require very different functioning abilities than bathing in the United States, such as having to walk half a mile to the nearest body of water. Roles of spouses and parents may also differ; some may be flexible and allow a variety of ways to fulfill the role expectations; others may be more fixed or rigid, requiring specific functional capacity that a person can no longer accomplish. Functioning is the linchpin for the individual's behavior, but the culture is what creates the expectations that need to be addressed by that behavior.

A further word about the idea of functioning: Functioning is a versatile concept and can be applied at many levels of activity, from the simplest to the most complex. Therefore, "functioning" is one of the more ambiguous terms used in the discussion of disability, one that contributes to the confusion that is created by the use of language to represent these concepts and their translation into measurement. A social worker and a physician may both point to the person's functioning as being disabling because it is restrictive or restricted. The physician is talking about the functioning of the person's kidneys that do not effectively cleanse the poisons from the body and are thus restrictive of the body's appropriate level of functioning (meaning the interaction of the various organ systems to maintain capacity); the social worker may be talking about the person's fatigue (caused by the poor kidney function) that influences the person's overall level of functioning. This functioning can be understood in terms of a person's willful actions: how far a person can walk, whether he or she can climb steps to reach an objective, or the clarity of his or her mind to allow good decision making. One additional use of the term "functioning" involves how well the person can meet the expectations of the organization providing a job or the needs and expectations of a spouse or a child to fulfill the role of parent or spouse. This last form of functioning, at the role level, is discussed next.

The fourth conceptual element according to Nagi is the actual behavior that evolves from the circumstances in which people find themselves. This is a result of the interaction situation (shown in the adapted Nagi model and the IOM model) in which the cultural expectations of the role and the nature of the physical environment affect the person, forcing modifications in behavior in order to adjust to the situation. In Nagi's version, the meaning of disability is "a pattern of behavior that evolves in situations of long term or continued impairments that are associated with functional limitations" (Nagi, 1965, p. 103). In some instances, the cultural expectations for a person or the barriers of the physical environment will preclude reasonable involvement in the role activity (such as working at an auto plant), and the person's ultimate behavior is to withdraw from an impossibly difficult situation. In other instances, a person may find that personal help (either another person or an assistive device) will allow him or her to modify behavior in such a way as to continue to satisfy some of the most important elements of a role. The role is not sacrificed; it is adapted to allow for a person's continued participation. In this instance, the flexibility of the cultural expectations and the adaptability of the physical environment, along with the level of difficulty and type of functioning difficulty the person brings to the context, combine to define the situation. Change any part of that equation and the "disability" can become more or less pronounced.

Unlike the first two diagrams, which clearly locate the unit of measurement of the concept at the person level, the location of a measure of disability needs to be at the point at which the person and cultural expectation and physical environment intersect (see Figure A-7). This cannot be accommodated in as straightforward a manner as identifying a missing leg or that a person cannot climb a set of stairs. In a survey context, if the source of information is either a person or a representative proxy and one is unable to observe the situation of the interaction, few alternatives are left. One can take a functioning approach that indicates how much difficulty a person has in a specific role situation (similar to assessing the limitation in basic actions), or one can take a behavioral approach that tries to identify how behavior has been affected by the change in functional capacity imposed by the impairment and the context.

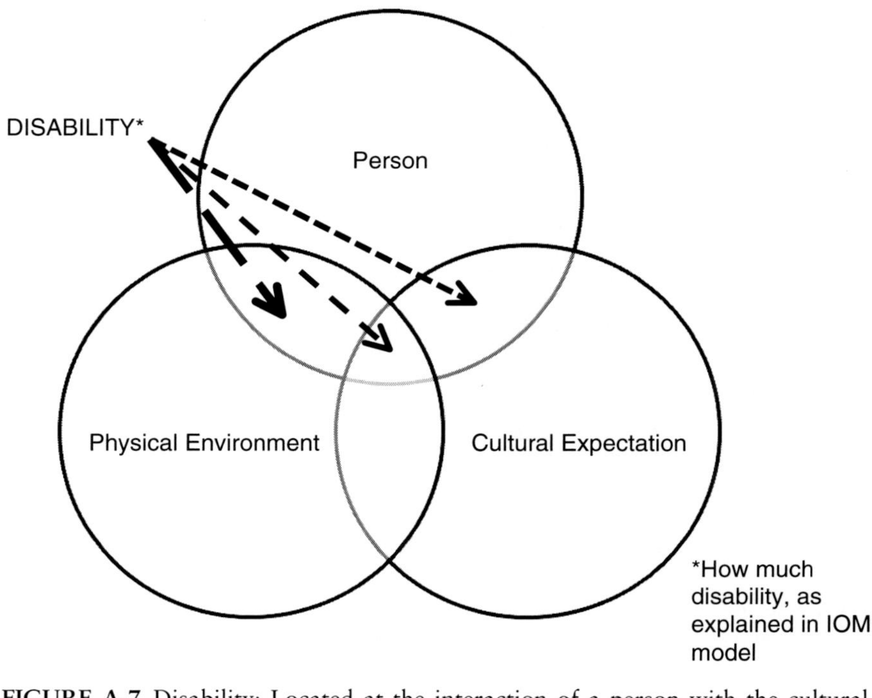

FIGURE A-7 Disability: Located at the interaction of a person with the cultural expectations or the physical environment or both.

There is one other aspect of this process that also needs to be considered. Nordenfelt (1995, 2003) has written about ability, opportunity, and choice as the three elements associated with a person's actions. Applied to a discussion of functioning, this would direct attention to the ability or capacity of the person to do—whether in a normative or nonnormative way—the physical, sensory, or cognitive actions called for; to be located in a place where such an action was possible and desirable; and to choose to do that action. An example may help to clarify this approach. Person A, who has limited use of the left side of her body, wants to know the weather outside. There is a window in the opposite wall from where she sits and a cane next to her chair. The person can choose to move across the room to observe the weather from the window or not. In this case, the will or lack of will of the person influences the functioning. If the limitations were different, more severe or encompassing both legs, the action would not have been possible with what was at hand. Finally, if there were no window or there was no cane available, there would have been no opportunity to accomplish the action. The choice element that Nordenfelt raises is not a consideration in any of the models, nor is it measured in current national surveys.

Purpose of Measurement

Another important background consideration related to the choice of concepts to operationalize during data collection is the purpose of the data collection, that is, what aspect of disability is the focus of concern. Because the data collected in national surveys are dictated by the orientation of the agency collecting the data—labor, health, or aging—the expertise of the agency colors and influences exactly what data are included and excluded and what aspects of the disablement process are seen as primary. When faced with making decisions about which component of a multidimensional concept is to be used to represent that concept as a data element, it is necessary to clearly understand which element of the concept is relevant to how the data are to be used (not something that is given a lot of conscious thought in many instances).

If the purpose of the data is relatively straightforward, such as the monitoring of trends in prevalence estimates of disability over time, the measurement can be relatively simple. Monitoring changes in national estimates of persons with limitations in ADLs and IADLs is an example: The data are identifying a single component of the disablement process and examining its fluctuation over a particular time or among groups. The only concern is to keep the question the same to allow for an accurate measure of trends over time. However, as my colleagues and I have noted there are multiple purposes for measurement of disability in national surveys, and that, combined with the multidimensionality of the concept, makes the development of survey questions more complicated (Altman et al., 2006). Fulfilling a specific purpose for collecting disability data may require different dimensions of the disability concept; all measures will not satisfy all uses.

In addition to monitoring trends, another purpose of disability measures in national surveys is needs assessment, covering a wide range of service provision, such as resource distribution, rehabilitation services, housing, transportation, and provision of long-term care (Altman et al., 2006). The bulk of data collection that has addressed service needs of the aging has concentrated, for example, on data about ADLs and IADLs limitations, which reflect whether the person has the ability to maintain independence and management of self-care. Prevalence estimates of those with difficulty maintaining or inability to maintain this independence in self-care have been used in analyses providing estimates of persons in immediate need of services or projecting possible future needs to influence policy decisions about long-term care. The needs assessment purpose has been a driving force in collecting disability data since disability became a formal administrative category in programs such as Social Security and vocational rehabilitation.

A new type of legislation that extends civil rights to persons with disabilities has created the necessity for a radically different approach to disability measurement. Civil rights legislation generates the requirement to monitor the equalization of opportunity of the population in question. Rather than being concerned with providing services for those most in need, this type of legislation requires that any person at risk of disability be included in the monitored population and that all types of access and social involvement be included. The focus of a contemporary civil rights perspective is located both at an activity and a participation level, to use ICF terms. However, to understand the mechanisms involved that facilitate or restrict the level of participation for an individual with impairments and concomitant functioning limitations, we must step back from measuring at the level of participation and focus on anyone who exhibits functional limitations (Madans et al., 2004). The newly proposed and tested questions in the American Community Survey (ACS; see Annex) were developed to allow for monitoring equalization of opportunity and predicated on the reasoning just discussed (Brault et al., 2007).

Operationalization of Concepts: Translation to Measures

The real challenge of measurement comes in translating the conceptual component of the model into an operational element that will accomplish the purpose of the data collection and capture the aspects of the situation that are relevant to address the issues. However, we narrow our conceptual elements when we are forced to operationalize them in the real world. Time, money, and respondent burden all limit the number of questions that can be asked and precipitate very hard choices, choices that will circumscribe full understanding of the process of disability.

Figure A-8 demonstrates what happens when one moves from the conceptual component of a model to a survey measure. Theoretical models, while abstractions of the real world that are useful for identifying the key elements and their relationships in broad-based terms without the confusion or detail of a specific empirical situation, are not practical for generating data (Altman, 2001a). They provide working and thinking space to focus on the basic ideas without getting caught up in details. However, to use the theoretical perspective, it needs to be applied to the empirical world and translated into questions that provide the data. An example of this process and how it narrows the focus of the theoretical concept would be the operationalization of the concept of functioning as identified in the Nagi and IOM models.

Functioning, which can refer to any aspect of an individual's physical, mental, or emotional actions, from walking down a road to hearing a conversation to putting dishes on a high shelf in a closet, is reduced to the number of elements about which there is space to ask. However, whatever selections are made, they come to represent that broader concept. If there is enough space to ask about a range of basic actions, such as walking, lifting, seeing, hearing, and communicating, there probably is not enough space

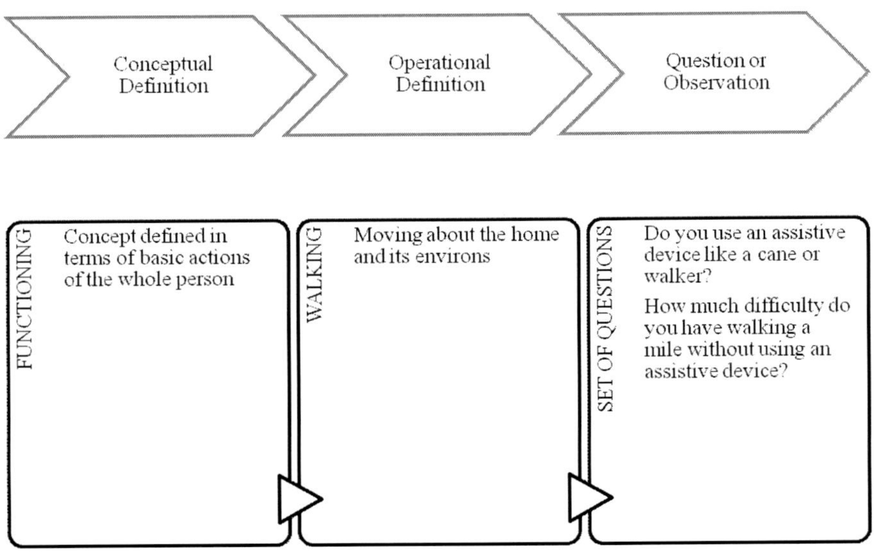

FIGURE A-8 Moving from concept to question using functioning as an example.

to ask about the details of the functioning, such as how far a person can walk, or how long, whether he or she can walk up and down steps or over rough terrain or stand for long periods of time. The choices made in this circumstance define the concept as it will be measured and used in analysis, not necessarily as it has been conceived in a general model. One hopes the choices will at least reflect the needs of the data collection related to the purpose of that data collection; however, they frequently do not. If the purpose is concern about aging people's ability to maintain their independence in the community, it is best to consider what functioning abilities are most related to that objective. An individual's environment is another equally important facet involved in such an objective, which I discuss below.

The ICF scheme provides a useful tool to help clarify the effects of operationalization on the development of measures to examine the disability process. Consider the chapter on communication in the ICF manual as a model (World Health Organization, 2001a). In understanding the Nagi, IOM, and ICF models, we recognize that communication is an area of functioning that is important to a person's participation in the role interactions that make up human existence. In the domain as outlined in ICF (World Health Organization, 2001a), numerous aspects of the communication process are identified, including the major components of communication, such as receiving spoken messages; receiving nonverbal messages; receiving written messages; speaking, producing nonverbal, sign language or written messages, and starting and sustaining conversation. This allows at least 9 or 10 areas through which one can operationalize the concept with the use of one or more questions, each providing a partial view of the functioning ability of the respondent in the area of communication. Each area of functioning—ambulation, using the arms and hands, hearing, seeing, learning, making decisions, and much more—would add yet another set of possible descriptive elements to be considered in the process of representing the concept. The simplicity of the theoretical definition gets lost in the specifics of the measurement potential. So although researchers may be starting from the same definition of functioning, their measurement and results may be as different as night and day.

Table A-1 shows the conceptual components of the various models, their definitions, and a selection of the operationalization possibilities. The table also indicates measurement possibilities for the various operationalizations and indicates the most commonly used measures. A comparison of the conceptual definitions and the common measures shows dramatically how far from the theoretical understanding of disability the actual measurement process brings us. By examining the information in the table, we see that not only are there many possible sources of indicators of functional limitation, disability, activity, or participation, but there are also several

TABLE A-1 Translation of Concept to Measure

Model Definitions and Examples	Pathology	Impairment	Functional Limitation
Nagi model	Mobilization of body's defenses and coping mechanisms as a result of infection, injury, or other etiology	Anatomical or physiological abnormalities and losses	Manifestation at the level of the organism (or person) as a whole of limitations of physical, sensory, cognitive, or emotional function
IOM model	Molecular, cellular, or tissue changes caused by disease, infection, trauma, congenital condition, or other factors	Losses of mental, anatomical, or physiological structure or function	An inability or hampered ability to perform a specific physical or mental task, such as climb a flight of stairs
ICF model	Health condition (disease or disorder)	Loss or abnormality in body structure, physiological function (including mental function). Refers to a significant variation from established statistical norms	An umbrella term for body functions and structures, activity and participation. Positive aspect of the interaction between an individual with a health condition and the person's contextual factors

Activity Limitation (ICF)	Disability (Nagi)	Participation Restriction	Environment
Part of disability	Social rather than organismic functioning—an inability or limitation in performing socially defined roles and tasks (Nagi, 1991). Also, a pattern of behavior that evolves in situations of long-term or continued impairments associated with functional limitations (Nagi, 1965)	Part of disability	Factors outside the person that contribute to shaping the disability, including the definition of the situation by others and their reaction and expectations, especially significant others. Also, the characteristics of the physical and social environment and the degree to which it is free from or encumbered with physical and cultural barriers (Nagi, 1991)
Part of disability	Same as Nagi (1991). The amount of disability a person experiences is a function of the interaction between the person and the environment	Part of disability	Represented as having two general categories, the social-psychological and the physical. Depicted as a three-dimensional mat
Difficulties an individual may have in executing task or action. It represents the individual perspective of functioning*	An umbrella term for impairments, activity limitations and participation restrictions. It denotes the negative aspects of the interaction of individual and contextual factors	Problems an individual may experience in involvement in life situations. It represents the societal perspective of functioning	All aspects of the external or extrinsic world that form the context of an individual's life that have an effect on a person's functioning

continued

TABLE A-1 Continued

Model Definitions and Examples	Pathology	Impairment	Functional Limitation
Operational definition examples	Cellular or tissue damage	Organ or organ system damage or anomalies	Organism basic actions or performance, includes complete range of human functioning abilities, such as seeing, speaking, walking
Measurement options examples	Diagnosed conditions or impairments, health care records, physical exams	Diagnosed conditions or impairments, health care records, physical exams	Communication, walking, seeing, hearing; various cognitive functions
Most commonly used measures	Diagnosed conditions or impairments	Diagnosed conditions or impairments	Physical, sensory, and cognitive functioning

NOTE: ICF = International Classification of Functioning, Disability and Health; IOM = Institute of Medicine.
*Contains combined ICF definitions of activity and activity limitation.

types or levels of measurement. The following discussion identifies those different types of measurement.

Types or Levels of Measurement

Figure A-9 shows the variety of levels of measurement that can provide information about disability or activities and participation, depending on the model being used by a researcher. Unlike impairment or body structure and function categories, which can be represented as present or not present and expanded by an indication of the level of severity when the problem is present, the diversity of human activity is much more difficult to capture than a physical attribute and exists at several levels of complexity. Some of the nature of that complexity is demonstrated in the figure.[4]

In contrast to the measurement of impairment, which is most frequently

[4]See the Annex for definitions.

Activity Limitation (ICF)	Disability (Nagi)	Participation Restriction	Environment
Includes actions as well as a range of tasks, from simple walking to more complicated activities, like using transportation	Household, work, civic, or community tasks performed in all the social roles in which a person can participate, including spouse, parent, worker, citizen, congregation member, friend	Includes tasks associated with specific major role designations, such as education, work, or community membership	Social and physical contexts in all social situations, including built structure, organizational policies and structure, cultural laws and norms
All types of mobility, communication, learning, self-care tasks, domestic tasks, relationships	Behaviors or functioning associated with self-care, working, maintaining a household, family roles, or civic roles	Performance of tasks required to engage in school, work, or economic, civic, or community venues	Assistive devices, home characteristics, transportation characteristics, building characteristics, climate and topography
Physical functioning, self-care, and domestic tasks	Self-care, household tasks, work, and leisure activity	Involvement in or limitation in school or work	Assistive device use; household characteristics

represented in survey data by self-report of a physician's diagnosis, there are extensive sets of questions and a variety of approaches to questions that reflect the hierarchy of complexity associated with the disability process. At least four additional levels of measurement when applied to the experience of the individual with an impairment have been elaborated (Madans et al., 2004). Using the ICF classification, Chapter 4, to provide a more concrete example of possible areas of operationalization, one gets a sense of the different levels that need to be reflected in measurement. At a very basic level, Chapter 4 provides a detailed classification of movement, in which categories are defined on the basis of movements designed to change body position or location, such as bending, standing, sitting, etc. This domain also includes categories associated with carrying objects, using the hands for grasping and propelling the body on land or in the water. Except for the last category, these activities can be conceptualized as measuring basic *willful* actions, or movement of bodily parts in a deliberate, intentional

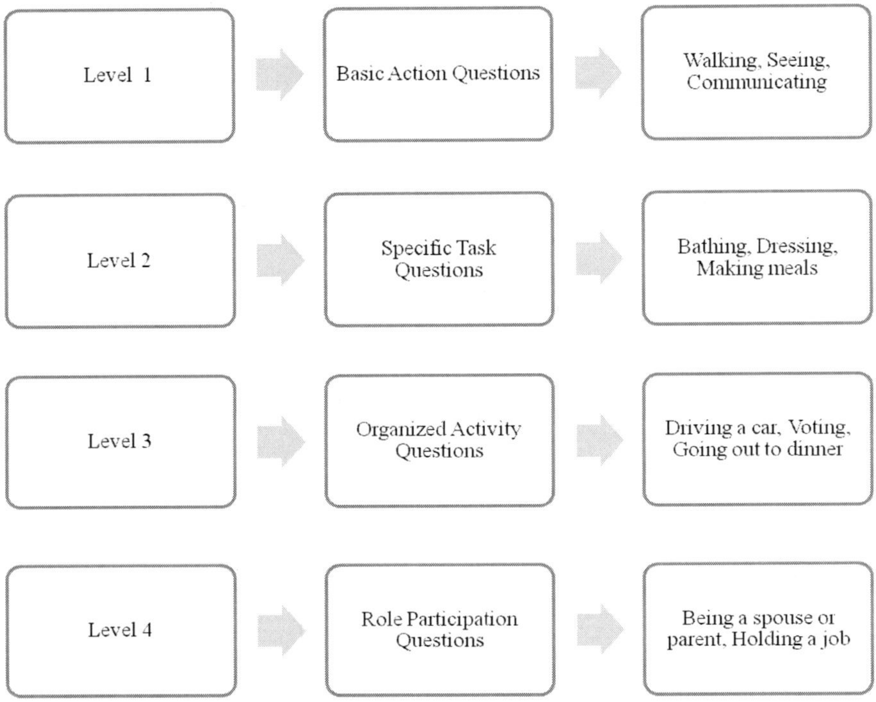

FIGURE A-9 Four types or levels of measurement. See the Annex for more detailed definitions.

process to accomplish a single physical objective, such as threading a needle or traversing a room (Nordenfelt, 2003). A more complicated level of measurement that goes beyond the coordination of bodily movement in a holistic way is represented by a *specific task*, in which an individual is motivated to combine physical movement, intellectual activity, and the use of assistive devices in an organized process in order to reach a specific goal. So for example, an individual would use what body functions and intellectual capacity are available to him or her in order to dress to go out or to prepare a meal.

The most complicated level of mobility for an individual is involvement in an *organized activity*, focused on combining basic individual actions with multiple tasks in order to accomplish one or more activities associated with a specific role. It represents the completion of a combination of ongoing basic willful actions and multiple specific tasks necessary to engage in behavioral elements of an ongoing role. An identifying quality of an organized activity is that it almost always involves not only the person in question, but also negotiating interaction with other people or the management of behaviors that tacitly adhere to accepted customs of the social system that the culture has developed. Chapters 6 and 7 and, particularly, Chapters 8 and 9 of the ICF (World Health Organization, 2001a) give extensive examples of organized activities that are representations of various components of involvement through completion of complex task elements associated with various roles. The area of mobility can provide an example, in the management of a motor vehicle or the use of public transportation. Personal use of a motor vehicle requires coordination of visual and physical and cognitive actions to manipulate the machine, the necessary attention to laws about using the roadways (speed limits, stop signs, rights of way, etc.) to move the vehicle on the roadway, and geographic familiarity with the area in order to maneuver the vehicle from point A to point B. Driving a car or other vehicle is an important organized activity, because it is related to a variety of roles and may even be a central organized activity for the work role of a truck or bus driver.

Ascertaining that an individual is capable of completing a basic willful action gives no information about his or her ability to engage in an organized activity, but involvement in an organized activity does imply that the person is capable of some form of basic willful action and specific task completion, and so the levels of measurement identified here do reflect a crude hierarchy. Whether the basic actions or the tasks are carried out with the aid of assistive devices or accomplished in an unusual way (using one's feet instead of hands) is irrelevant to the measurement of that element of participation; however, such detail may provide additional information on the person's functioning or the environmental context.

The final level of measurement is an indication of *participation* in a cultural role. This can be accomplished through either a functioning approach or a behavioral approach to asking questions. A functioning approach would use indicators of limitations or inability to participate in the role, or alternatively, indicators of level of difficulty in performance of the role. An example would be a teacher who answers a question about limitations in work as limited because she can no longer stand before a classroom to lecture, but who, with the help of her employer, arranged to work with students on a one-to-one basis. Alternatively, a respondent may indicate being completely unable to fulfill certain of the main activities associated with a role. An example of this would be a teacher who has lost the ability to communicate verbally or otherwise, who can no longer interact with students in an instructive way. An alternative functional approach would have the teacher indicate the degree of difficulty experienced in performing a variety of tasks and organized activities required of the teaching role, including an indicator that it cannot be done.

A behavioral approach to participation seeks to understand changes in behavior that can be attributed to changes in capacity (or basic action functioning). Questions asked from this approach would seek to discover if a person was involved with this particular role or activity prior to the onset of the impairment or the functioning limitation and whether involvement has changed in any notable way[5] (instead of answering the functioning component above as being limited, the change in the way the person teaches could be captured). In some instances, questions have sought to discover if the role behavior had increased, decreased, or remained the same (Ferron, 1981). Other approaches could identify situational changes that allowed for maintenance of the role, thus incorporating some of the environmental factors that are presently so scarce.

SURVEY-RELATED MEASUREMENT OF DISABILITY

An examination of the large, general national surveys and the surveys directed toward the aging population shows a remarkable consistency in representations of the very different conceptual elements included. Moreover, they highlight the gaps in information collected when considered in light of all the possibilities discussed above. Tables A-2 and A-3 show the relationship of questions on disability to the theoretical concepts from current models found in general national surveys (see Table A-2) and specialty surveys on aging (see Table A-3). The following discussion examines the various representations of the theoretical concepts, the levels or types of measurement, and the specific functioning or role participation emphases included in each of the surveys; the discussion also identifies the gaps in information contained in the two sets of surveys.

Theoretical Concepts Represented in Survey Questions

Medical and Chronic Conditions

Although the theoretical models of disability are careful to identify and differentiate the concepts of pathology and impairments in the context of health, disease, injury, and congenital anomaly, the survey approach to identifying that element of disability consistently focuses on either identifying a specific health condition or framing questions about other conceptual areas in the context of a physical, mental, or emotional health problem. Except for some questions included in the Women's Health and Aging

[5]The Social Security Administration used questions of this nature in a series of surveys in 1972 and 1974. They combined to provide a scale that very effectively represented the net result of having an impairment and functional limitation in terms of participation at the time of the survey (Altman, 1984).

TABLE A-2 Theoretical Content of Disability Questions Contained in General Social Surveys

Survey Question Content	Pathology or Condition: Impairment	Functioning: Basic Actions[a]	Disability: Adaptive Behavior[b]	Participation: Role Functioning[c]	Environment[d]
National Health Interview Survey					
Family level[e]					
Limited in any way		X [1]	X	X	
Remembering or periods of confusion		X [1]			
General ADLs—need help				X [6]	
General IADLs				X [1]	
Working				X [2]	
Conditions	X [list]				
Use assistive device					X [1]
Administrative measures[f]	Reflects adjudication as eligible for disability benefits based on medical certification required by program. Related to identification of condition or impairment. [2]				
Sample adult					
Nagi functioning—difficulty level		X [9]			
Conditions causing difficulty	X [list]				
Seeing		X [2]			
Hearing		X [2]			
Emotional functioning[g]		X [7]			
Social/leisure activities				X [3]	
Use of assistive device					X [1]

continued

TABLE A-2 Continued

Survey Question Content	Pathology or Condition: Impairment	Functioning: Basic Actions[a]	Disability: Adaptive Behavior[b]	Participation: Role Functioning[c]	Environment[d]
NHANES					
Sample person					
Nagi functioning		X [8]			
General ADLs—difficulty				X [5]	
General IADLs—difficulty				X [5]	
Seeing		X [9]			
Hearing		X [7]			
Cognition		X [1]			
Use of assistive device					X [3]
Conditions causing difficulty[h]	X [list]				
Work/housework functioning				X [3]	
Work/housework behaviors[i]			X [8]		
Social/leisure activities				X [3]	
Social/leisure activity behavior[j]			*X [6]*		
Comparison with 10 years ago			*X [1]*		
MEPS[k] 2006					
Family sample					
General ADLs[l]—receives help/time			X [3]		
General IADLs—receives help/time			X [3]		
Use of devices					X [1]
Nagi functioning questions		X [8]			

Work limitations				X [2]
Social activity limitations				X [1]
Cognitive functioning		X [3]		
Seeing		X [5]		
Hearing		X [5]		
Condition	X [list]			
Conditions associated with health visits	X			
Self-administered Questionnaire				
SF-12 questions[m] [8]	The questions in this self-administered question set do not lend themselves to easy separation into the categories that have been developed here.			
Frequency of feeling states[n]		X [4]		
Health today[o] [5]				
SIPP—2004				
Household sample				
Nagi functioning questions[p]		X [17]		
General ADLs—difficulty				X [7]
General ADLs—need help				X [7]
General IADLs—difficulty				X [5]
General IADLs—need help				X [5]

continued

TABLE A-2 Continued

Survey Question Content	Pathology or Condition: Impairment	Functioning: Basic Actions[a]	Disability: Adaptive Behavior[b]	Participation: Role Functioning[c]	Environment[d]
Seeing		X [2]			
Hearing		X [2]			
Speaking		X [2]			
Use of assistive devices					X [1]
Conditions[q]	X				
Cognitive and emotional health conditions[r]	X [5]				
Emotional functioning[s]		X [5]			
Work/housework				X [4]	

NOTE: LSOA = Longitudinal Survey of Aging, MEPS = Medical Expenditures Panel Survey, NHANES = National Health and Nutrition Examination Survey, NHIS = National Health Interview Survey, SIPP = Survey of Income and Program Participation.

[a]Basic actions represent deliberate willful functioning of the whole person.

[b]Disability as it is described in Nagi as behavior that develops over a period in a situation of long-term impairment associated with functional limitation within the individual's social and physical environment. The disability is not the functioning, but the behavior that evolves.

[c]Reflects functioning in role situation, from maintaining independence to fulfilling expectations as a parent, spouse, employee, etc.

[d]Represents the two environmental components of cultural expectations and social and physical environment.

[e]These questions are asked of the designated family respondent about the whole family and are considered the core questions, while a more extended set of questions are asked about a sampled adult and a sampled child within the family.

[f]Collects information on Supplemental Security Income and Social Security Disability Insurance.

[g]Measures reflecting the Kessler 6 can represent psychological distress and when teamed with a question about whether the symptoms interfere with daily activities can represent difficulties with emotional functioning.

[h]Extensive questions about conditions and their treatment.

[i]In NHANES there are a series of questions about changes in work and housework behavior that are different from asking about having limitations in type or amount of work, which is a functioning measure.

[j]NHANES contains a set of four questions on the frequency of social behaviors that lends itself to being used as a social behavior indicator if a comparison element is added.

[k]MEPS is a longitudinal survey over a period of 2 years. The sample is drawn from the NHIS, so that 3 years of tracking are possible.

[l]Screeners ask about receiving help or supervision, which reflects on behavior rather than functioning.

[m]Difficult to categorize by the conceptual components adopted here. Mix functioning-, behavioral-, participation-, and symptom-focused questions with health status and consider it general health.

[n]Provides frequency of a particular feeling state in the past 4 weeks as opposed to the level of difficulty such a feeling state may create—peripherally useful.

[o]Combines indicators of functioning (physical), participation functioning (self-care and usual activities), and pain symptom. Intended as a combination measure and cannot be attributed to any one category.

[p]Does not necessarily mean 17 different items of functioning, but rather is tied to providing yes/no answers so that subsequent questions ask about levels of difficulty.

[q]After asking about functioning, sensory, and ADL and IADL problems, this survey asks about the conditions that cause the health problem and which condition is considered the main reason for the difficulties. There is also a question to establish a start date for the problem.

[r]There is also a question that asks about the presence of five different types of conditions that can cause mental health or cognitive problems.

[s]Survey contains three questions that could be interpreted to represent emotional functioning associated with getting along with people, ability to concentrate, and coping with stress that interferes with everyday activities.

TABLE A-3 Theoretical Content of Disability Questions Among Surveys on Aging

Survey Question Content	Pathology/ Condition: Impairment	Functioning: Basic Actions[a]	Disability: Adaptive Behavior[b]	Participation: Role Functioning[c]	Environment[d]
LSOA—2000					
Nagi functioning questions—any difficulty		X [7]			
Nagi functioning questions—level of difficulty		X [7]			
ADLs—any difficulty				X [6]	
ADLs—amount of difficulty				X [6]	
Special equipment used			X [1]		
Type of special equipment					X [6]
ADLs—receive help or need more help			X [6]		
Need help with ADLs				X [6]	
IADLs—any difficulty				X [7]	
IADLs—amount of difficulty				X [7]	
IADLs—receive help or need more help			X [7]		
IADLs—need help				X [7]	
Conditions	X [list]				
Vision[e]	X [3]				
Hearing	X [4]				
Cognitive functioning[f]	X [test]				
Mental health	X [3]				
Social behavior[g]				*X [12]*	
Leaving home[h]				*X [2]*	
Housing questions[i]					X [5]

NLTCS—2004					
Screener					
ADLs		X [9]			
IADLs		X [7]			
Conditions	X [list]				
Impairments	X [list]				
Conditions as main cause of ADLs	X				
ADLs help[j]			X [6]		
IADLs help[k]			X [9]		
Use of special equipment[l]					X [8]
Cognitive functioning	X [test]				
Mental health Treatment related[m]	X [5]				
Emotional functioning		X [5]			
Nagi functioning questions		X [8]			
Special equipment outside					X [3]
Seeing		X			
Hearing		X [2]			
Speech		X [2]			
Social behaviors[n]			*X [9]*		
Housing characteristics					X [10]
HRS—2008					
Nagi functioning questions		X			
General ADLs—difficulty				X [6]	
General ADLs—get help			X [6]		

continued

TABLE A-3 Continued

Survey Question Content	Pathology/ Condition: Impairment	Functioning: Basic Actions[a]	Disability: Adaptive Behavior[b]	Participation: Role Functioning[c]	Environment[d]
General IADLs—difficulty				X [5]	
General IADLs—get help			X [5]		
Conditions					
Cause impairment	X				
List	X				
Mental health (9)		X			
Social behavior (3)			X How often		
Work[o]				X [4]	
Employer-provided accommodations					X[10]
Hearing		X [2]			
Seeing		X			
Cognition/memory[p]		X [test]			
Driving[q]			X [5] Measured at organized activity level		
Use of equipment[r] and what type					X [5]
WHAS—1995					
Screener					
Nagi functioning questions—any difficulty		X [5]			
Nagi functioning questions—how much difficulty		X [5]			

General ADLs—any difficulty				X [5]	
General ADLs—how much difficulty				X [5]	
General IADLs—any difficulty				X [5]	
General IADLs—how much difficulty				X [5]	
Conditions list	X [13]				
Cognitive functioning		X [test]			
Sampled person[s]					
Nagi functioning questions		X [8]			
Start date/main condition cause	X [8]				
Use of devices for functioning and frequency					X [6]
General ADLs				X [5]	
Do differently Get help			X [5]		
Use of devices					X [5]
General IADLs				X [8]	
Done less often or differently/help			X [8]		
Conditions[t]	X [list]				
Mental health[u]		X [36]			
Social behavior and frequency[v]				X [4]	
Seeing		X [5]			
Hearing		X [11]			
Quality of life (20)	Intervening concept in IOM model				

continued

NOTE: HRS = Health and Retirement Study, IOM = Institute of Medicine, LSOA = Longitudinal Survey of Aging, NLTCS = National Long-Term Care Survey, WHAS = Women's Health and Aging Study.

[a]Basic actions represent deliberate willful functioning of the whole person.

[b]Disability as it is described in Nagi as behavior that develops over a period in a situation of long-term impairment associated with functional limitation within the individual's social and physical environment. The disability is not the functioning, but the behavior that evolves.

[c]Reflects functioning in role situation, from maintaining independence to fulfilling expectations as a parent, spouse, employee, etc.

[d]Represents the two environmental components of cultural expectations and social and physical environment.

[e]Questions about vision and hearing asked about the body part function rather than the functioning of the whole person.

[f]Indicators of cognitive functioning based on a series of tests of the person's cognitive capacity, mainly memory, and orientation.

[g]Social behavior questions have the potential to provide insights into behavior changes associated with disability if altered somewhat.

[h]Questions about leaving home also provide an opportunity to identify behavior change associated with disability if questions are adapted to capture the necessary information.

[i]Provide some environmental information.

[j]ADL questions contain a whole series of information about use of help, supervision or standby help, use of assistive devices, how long help was needed, how frequently help was needed.

[k]Similar to ADL question in focus on help, assistive device use, and timing.

[l]Survey contains a variety of special equipment questions associated with ADL limitations and IADLs. These questions refer to activities inside the house. There is also information on assistive devices related to going outside, vision, and hearing in other locations throughout the survey. These are noted below.

[m]There are two questions about hospitalization and doctor use for mental health problems and three questions about the use of medication.

[n]Social behaviors if adapted to changes pre- and post-condition, impairment, or functioning onset would reflect disability behavior.

[o]Work questions focus on paid work and housework and identify limitations as well as being totally unable. However, a group of additional questions provides information about ability to work full- or part-time; doing the same or different work and the provision of some accessibility by the employer.

[p]Provides extensive set of questions to test memory, other areas of cognitive functioning scattered about (ability to follow a map), but emphasis on memory.

[q]Very interesting combination of questions to ascertain information about the organized activity of driving, including recent behavior.

[r]Asked in conjunction with ADL questions.

[s]Limited to women aged 65 and over.

[t]Condition list limited to 13 types of conditions, but detailed questions provide much more information about location of impairment or the nature of the symptoms.

[u]Extensive set of questions reflecting symptoms associated with depression and anxiety.

[v]Questions in other sections about bowling and card games may reflect social behavior, but are asked in terms of symptoms experienced when doing them.

Study (WHAS), the survey data represented in these tables do not allow a researcher to separate the condition-caused pathology from the impairment, nor are impairments identified, except possibly in the case of accidental injury or amputations for which some surveys indicate the location of the injury. Medical conditions serve as the catch-all for the pathology or impairment indicator and, except for the National Health and Nutrition Examination Survey (NHANES) and WHAS, in which some information is based on actual examination, responses are based on either self-reports or a self-reported indication that the condition was diagnosed by a physician. Surveys with links to administrative data—such as Medicare records in the National Health Interview Survey (NHIS) or physician follow-back for a subsample of the Medical Expenditures Panel Survey (MEPS)—can verify a diagnosis through the administrative record, but that is a factor that complicates not only the data collection process, but also the data programming task during analysis.

Conditions are most frequently collected by a bounded listing that emphasizes the most frequently occurring chronic diseases, such as heart disease, arthritis, cancer, and stroke.[6] In addition, in some surveys, such as NHIS and WHAS, after being asked the questions about basic action functioning and task topics, respondents are also asked to identify the condition or conditions that are considered to be the main cause of the limitation that has been identified. Although in many cases this may be an accurate representation of the cause of impairment and functioning difficulties, particularly when only one condition is involved, respondents frequently indicate the condition occurring most closely in time to the development of the limitation, even though the limitation may be a result of an earlier condition that has worsened, or, most likely, the accumulation of conditions. Unlike the other surveys considered here, MEPS also identifies conditions in addition to those on the list that are associated with all types of medical care use, physician visits, hospitalizations, medication use, and visits to other providers. This information can serve as a check on what may be the major conditions occurring during the data collection period that may also be associated with the limitations being reported. This material is particularly useful for picking up secondary conditions that can contribute to the worsening of functional limitations, as demonstrated by Rasch and colleagues (2008).

[6]On the MEPS and the Health and Retirement Study (HRS), see the health status sections for examples of the conditions considered important to capture.

Physical Functioning

Functioning, as identified in the Nagi and IOM models and partly identified by the activity concept of the ICF model, is represented in survey data by various versions of what some call the Nagi questions. The Nagi questions ask about physical functioning, associated primarily with movement and use of arms, legs, fingers, and the body as a whole. The original Nagi questions (Nagi, 1969), which were asked relative to a work context, were rather detailed in that they asked about lifting a series of weights ranging from 1–5 pounds to 50 pounds, asked about the left and right hands and arms separately, asked about sitting and standing, and separated the functional aspects of stooping, crouching, and kneeling. The Nagi questions found in the surveys under study are generally fewer and less specific and continue Nagi's omission of questions related to sensory functioning. They all include at least one question about walking, and the more elaborate ones ask about the ability to walk several different distances and up steps (it appears that going down is assumed to be less difficult). Some surveys, like NHIS, ask about nine different areas of functioning, including reaching overhead with the arms, bending and stooping, using the fingers to grasp, lifting a certain amount of weight (usually about 10 pounds), and pushing or pulling furniture. Others, such as WHAS, ask about five areas of functioning, but in somewhat more depth.

Cognitive and Emotional Functioning

Other elements of functioning that the surveys attempt to capture are cognitive functioning, predominantly the ability to remember; mental health status, which is assumed to provide some insight into emotional functioning; and sensory functioning—seeing and hearing. Most of the general national surveys are limited to one question about cognitive functioning. For a short set, MEPS seems to be best because it combines three different capacities: memory loss or confusion that interferes with daily activities, problems making decisions, and requiring supervision for safety. In the other general surveys, cognitive function is equated with memory problems, while in the aging-specific surveys, various forms of Mini-Mental tests are conducted. Although those Mini-Mental tests do identify deteriorating cognitive capacity, they do not necessarily indicate how the person functions in daily activities. In the same way, the indicators of emotional functioning are generally associated with either a depression scale or a depression and anxiety measure. In some instances, questions directed at other concepts provide clues as to level of emotional functioning, such as questions in WHAS about how well a person sleeps or experiences shakiness in a particular situation. However, unlike questions that ask about the amount of difficulty a respondent has with physical movement (which gives some sense of the level of functioning), mental health questions are yes/no recitations of symptoms that probably mask the relationship of the symptom to the basic actions that would be related to emotional functioning. The actual problems created by those symptoms, such as with decision making, safety, emotional control, or withdrawal, and the level of those problems are simply assumed on the basis of the level of symptomatology. At this time, understanding of how to measure emotional functioning lags behind measurement of other capacities.

Sensory Functioning

Measurement of sensory functioning, primarily seeing and hearing, is included in all the surveys reviewed for this paper, but the adequacy of the measures to indicate functioning, even functioning potential, is poor. For example, the two questions in the Survey of Income and Program Participation (SIPP) on vision ask about difficulty seeing the words and letters in ordinary newspaper print (even when wearing glasses); at best, they only separate people who are blind or unable to see newspaper print from all others. This is useful for identifying people with severe difficulty seeing, but it does not indicate the range of difficulty or identify those who may have difficulty seeing in the distance (necessary to drive) or those with problems seeing at night, and so forth. The vision examination included in NHANES may be the one exception to this criticism, but a medical exam also does not necessarily indicate how an individual with a vision impairment is functioning in an everyday situation. Hearing questions are also insufficient in many instances, identifying those who are deaf and those who use hearing aids, but without an indicator for all others. The issue of hearing aids is also a serious complication in questions about hearing because aids do not serve to ameliorate hearing problems as well as glasses facilitate seeing.

Communication

One other area of functioning—speech—is represented only in the SIPP data. Although far fewer people experience speech difficulties than vision and hearing difficulties, it is an important area among the aging population for people who have had strokes or have conditions such as multiple sclerosis or Parkinson's disease. WHAS is the only survey that collects data on the type of functioning problems related to hearing and speech impairments that elders experience if they have had a stroke; yet it does not collect extensive data on limitations in understanding incoming conversation, which is part of the interaction and participation equation.

Role Participation

There is a wide variety in the measurement of role participation on which the surveys collect data. In Tables A-2 and A-3, I use two columns (columns 4 and 5) to represent that area, one reflecting the behavioral interpretation of participation that comes out of the early Nagi work (Nagi, 1965) and the other reflecting the functional approaches to participation found in the IOM and ICF models. The variety I speak about above, however, is not in the role participation topics; rather, it is associated with the wording, answer categories, and selection of self-care task items in the measures. All of these surveys, particularly the aging surveys, have extensive measures of self-care tasks, the ADLs and IADLs originally identified by Katz et al. (1963). These measures focus on the ability of individuals to maintain independence in elemental self-care areas, such as bathing and dressing, and in more complicated instrumental areas, such as taking care of meals and maintaining a household. This orientation to what are important, but minimal, self-care tasks comes out of an earlier era when disability was understood as dependence and public policy addressed such dependence with various forms of care provision, including institutions. At that time, dependence, a form of nonfunctioning, was seen as an end in itself, rather than as an obstacle to participation.

Data on ADLs, in particular, are collected not only about the existence of difficulty, but also about the amount of difficulty experienced, the use of help, the need for help, and the use of assistive devices to accomplish the task. The Longitudinal Survey of Aging (LSOA), the National Long-Term Care Survey (NLTCS), and WHAS are actually quite thorough in that regard. Among the general national surveys, only SIPP has multiple questions on ADLs and IADLs; the other general surveys have fewer questions and much less detail. The aging surveys, in contrast, ask a minimum of five questions about self-care and at least five questions about instrumental activities, such as making a meal, shopping, or managing medications or money. Questions that ask about difficulties or a need for help or supervision with ADL or IADL tasks are considered to reflect a functioning approach to role participation, while questions that indicate the use of help are interpreted as indicating task behavior on the part of the respondent; they are classified in Tables A-2 and A-3 accordingly. This is the one area of disability questions that has been examined more closely and whose standardization has at least been attempted, if not accomplished (Freedman et al., 2004).

Two other kinds of questions about role participation are included in most surveys. One is about functioning in the work role (and HRS also provides extensive data on behavior in the work role), and the other is social behaviors or role functioning. The work questions refer mainly to paid employment, although sometimes to housework as well. The social role questions, usually three or four, generally ask about difficulties in social activities, such as attending church or concerts; going out to movies or sporting events; and having leisure activities, such as reading, watching television, or listening to music. A very interesting aspect of my review was the discovery of several sections of measurement that with a little tweaking would provide some very useful data on role behavior, not just functioning (these sections are italicized in Tables A-2 and A-3). An example is the set of questions about social activity in LSOA. There are eight different items that identify eight different types of current social participation. However, there is no indication of how this behavior may have changed since the onset of the functioning difficulty. Increase or decrease in the activity, as well as the addition of new activities, would provide valuable information that could document the losses or gains in participation associated with functioning problems. Other nonstandard questions already in LSOA, NLTCS, and WHAS have potential to increase data in a useful way with very little change (see the Annex for suggestions).

The one final theoretical concept about disability included in Tables A-2 and A-3 is the area of environment. Measures accounting for environmental factors all appear to be restricted to the use of assistive devices and the kind of devices in use. LSOA has some minimal information about the individual's home, and HRS has a fairly detailed set of questions about accommodations made in the workplace by a person's employer. The latter set could usefully be borrowed by the Current Population Survey (CPS) to address this issue for all ages.

Levels of Measurement

The dominant levels of measurement of functioning found in the surveys I reviewed are level 1 and level 2, measures of basic actions or functioning of the person and measures of specific tasks, primarily those associated with self-care and independence. Although the question included in most IADL measures on going shopping can be considered a question about organized activity, there is very little detail considered other than accomplishing the task. Except for the one set of questions about driving in NLTCS, there are really no other questions about such organized activity as using public transportation, acquiring a job, attending college, traveling to visit friends or relatives outside the local area, taking vacations, voting, and many other kinds of activity. The level of measurement of role participation is also very limited. Questions about work experience simply reflect the people's assessment of their own ability to work, with no indication about what personal functional limitations or environmental access circumstances contribute to the answer. Information about other role involvement can be gained through answers to questions about marital status, parental status, and educational status, but little more. Questions about community roles (civic roles, such as voting or organizational membership in churches, clubs, or other groups) could well be included in surveys for the full population and would give a more rounded picture of life than is now available.

STRENGTHS AND WEAKNESSES OF CURRENT SURVEYS

Depending on the perspective one takes, the disability data available today, with which we analyze important related policy questions, is either remarkably extensive or too narrowly conceived. Actually it is both.

Although the collection of information on impairment dates as far back as 1830 in the U.S. Census, it is a fairly recent development to have multiple questions about disability available in national databases. In 1936 the U.S. Public Health Service fielded a National Health Survey that asked about a respondent's health and use of the health care system. A prominent component of the survey questions was the section on other handicapping disease or condition, which focused on recording all handicapping diseases or conditions for every member in the household. In addition to recording the start date of conditions and doctor use, the section allowed for collection of information about whether the condition "interferes with job" (Perrott et al., 1939). The purpose of those questions was to inform the public health sponsors about the health of the nation and the types of prevalent illnesses or diseases, as well as to identify the impairments (disabling or not disabling, although disabling was not defined). Exploration of disability other than accounting for the various conditions that caused it was beyond the purview of the survey.

In a similar fashion, NHIS, which started in 1957, had among its first publications a report on *Impairments by Type, Sex, and Age* (National Center for Health Statistics, 1959). The data were collected on an extensive list of impairing conditions, and prevalence rates were published. The concept of impairment was still included within illness and not explored as a precursor to "disability."

Observing the current set of surveys from these earlier vantage points, the data available in the surveys reviewed here have broadened to encompass representation of almost all areas of the more recent conceptualizations of disability and include impressive detail about individual functioning, in most instances, as well as information about maintaining independence, using help, and maintaining social ties.

Of the two forms of surveys reviewed here, general national surveys and specialized aging surveys, each group has important strengths. Among the general national surveys, the best coverage is for the conceptual area of individual functioning. Although all areas of physical, mental, and emotional functioning are not covered (speech is one notable area missing in all but SIPP), the areas of functioning that account for most functioning limitations are covered, including difficulty walking, using arms and hands, and sensory difficulties. As mentioned above, the questions about seeing, hearing, emotional difficulties, and cognitive limitations could be improved and expanded to more directly reflect actual functioning, but there are at least representations of those functioning areas. It would be particularly useful to include additional aspects most likely to result in organized activity or role participation limitations, such as cognitive and emotional functioning, so that the relationships could be investigated. Another strength of the general surveys is that they include adult respondents of all ages, so that one can examine the continuum of disability across a wide age spectrum and possibly observe patterns of relationships that are not available in a more age-restricted sample.

Weaknesses of the data relative to theory include very limited measures of role participation and no measures of environment except for the use of assistive devices. The use of a functioning limitation approach to role participation is somewhat restrictive in that it tells us only how respondents perceive their relationship to a specific area of participation, such as limited ability or inability to work. A behavioral measure of role participation, in contrast, would be more conducive to understanding the interaction process that creates disability. It would provide an approach that captures the nature of the change in role behavior and would be more useful for measuring the social, cultural, and physical environmental elements that are related to behavior changes.

The information from the aging surveys (and SIPP to a more limited extent) about self-care tasks is particularly detailed and useful for understanding the more severe areas of deterioration associated with aging and multiple impairing conditions. It is particularly useful for estimating the need for care services for this population, since in most cases the data allow researchers to differentiate between having difficulty with self-care tasks, getting help with those tasks, and in some cases needing help with those tasks. There is somewhat greater attention to role behaviors in the aging surveys, rather than just role functioning, and several of the aging surveys have question sets that, if tweaked slightly, would provide information not only about current behavior, but also about behavior change. This would be an important clue to the disability process or the development of "disability." One or two of these surveys do include environmental measures associated with the home setting, better than is done in the general surveys, but it is a small start in comparison to what is needed. And as with the general surveys, the aging surveys contained almost a total lack of data at the level of organized activity, which would inform about the activities in which role participation breaks down. In most instances, people do not give up social roles entirely; rather, they change the way they participate in them. Information about those activities for which there is change is useful for intervention purposes, for making things more accessible, or for encouraging greater participation. One final weakness of the aging surveys is the failure to recognize the cumulative effect of the disability process associated with aging. This process does not necessarily begin at age 65 and should be examined from an earlier point, such as 45 or 50, particularly if prevention is a goal.

Both sets of surveys have very similar data gaps. Except for the classic reference to going shopping found in IADL questions and the questions directed toward driving in NLTCS, none of these surveys provides information about organized activities associated with role participation. There are also very limited measurements of work roles and social role functioning. Given the emphasis on a functioning limitation evaluation of role participation rather than an examination of role behavior and how it might have changed since the onset of functional limitations associated with health problems, there is very little information about the disablement *process*. The outcome is known in a very narrow area, but the process of giving up actions or tasks as functioning deteriorates is not known, nor are the *choices* people make between or among role activities. As an example, people may accept help (if it is available) with ADLs and IADLs in order to have more time to engage in the tasks and activities that are meaningful to them in one or more roles. In the new conceptual orientations toward disability, the independence–dependence continuum is less important than participation and inclusion, even at the most dependent levels of functioning. There are now two policy issues related to this population: the cost and organization of care related to dependence and a mandate to facilitate inclusion and participation. The available data and research tell us very little about how to accomplish the latter, primarily because the data are not available, but also because data collection activities and research analyses are still operating under the old paradigm.

What the Measures Capture

The richness of the data available to represent the disablement process is sometimes a source of confusion to those who use the various analyses reported in government publications and journal articles. Silverstein et al. (2005) and the National Council on Disability (2008) have complained strongly about the lack of consistency of prevalence estimates from the various surveys that muddy the waters when policy issues are at stake. To a layperson who does not differentiate an ADL or IADL measure from a physical or mental functioning measure, the vast difference in prevalence estimates that are produced is confusing. Even researchers in the field sometimes inaccurately assume that persons who indicate a bathing or dressing difficulty all fall within a population estimate of persons with positive responses to the Nagi questions on functioning. Although the intuitive image of the nesting of one set of measures within the others in a hierarchical manner probably suits the broad generalizations implied by the theoretical models, the real-life examination of data does not provide such a neat package, as both Adler and I have demonstrated empirically using two different data sets (Adler, 1989; Altman, 2001a).

When examining these different measures that reflect the conceptual elements of theoretical models of disability, it is important to keep in mind that the different measures not only reflect a shorthand for the conceptual elements, but also represent different populations in the big picture of disability. The prevalence estimates and percentages in Table A-4, taken from a recent chartbook about disability and health, show those striking differences. Taken together, the combined measures of disability available in NHIS identify more than 66 million adults or 31.4 percent of the

TABLE A-4 Population Estimates Based on Concepts Used to Measure Disability, Total Adult Population, and Population Ages 65 and Over

Disability Measure	Population Estimate All Adults[a]	Percent	Population Estimate Ages 65+[b]	Percent
Total	211,133,000	100.0	33,061,000	100.0
Any limitation measure	66,317,000	31.4	15,692,000	47.5
Basic action Difficulty[c]	62,338,000	29.5	—	—
Movement difficulty	45,903,000	21.7	11,448,000	34.6
Sensory difficulty	27,655,000	13.1	3,892,000	11.8
Cognitive difficulty[d]	5,876,000	2.8	—	—
Emotional difficulty[e]	6,487,000	3.1	692,000	2.1
Role Participation	30,097,000	14.3	—	—
Self-care	8,738,000	4.1	ADLs: 2,002,000	6.1
ADLs and IADLs			IADLs: 4,008,000	12.1
Social role limitation	14,599,000	6.9	—	—
Work limitation	24,548,000	11.6	8,110,000	24.5

[a]Data from Altman and Bernstein (2008) reflect cumulated NHIS data for 2001–2005.

[b]Data from Harris et al. (2005) reflect NHIS data for 2002.

[c]Measured in terms of physical, sensory, cognitive, and emotional functioning.

[d]Cognitive functioning measure is based on one question about memory and experiencing confusion.

[e]Emotional difficulty is based on a score on the K6 > 13 or an indication that the feeling interfered with daily activities.

noninstitutionalized adult population. By far the greatest proportion of that population, more than 62 million or (29.5 percent) were identified by responses to questions about basic action functioning (physical, sensory, cognitive, and emotional); almost 46 million (21.7 percent) indicated some difficulty with the physical functions represented by the Nagi questions. Basic action difficulties represent other areas of functioning as well: 27.6 million (13.1 percent) indicate sensory difficulties; 6.5 million (3.1 percent) indicate emotional difficulties; and 5.9 million (2.8 percent) have cognitive problems, at least with memory and confusion. These total only 62 million because almost 19 million have limitations in more than one area of functioning.

A little less than half of the total population identified by limitations in basic actions were identified by the other commonly used measures, task or role limitation: 30 million (14.3 percent) reported a limitation in self-care, work, or social roles. Within the group with role limitations, 24.5 million (11.6 percent) reported work limitations; 14.6 million (6.9 percent) reported social role limitation, and 8.7 million (4.1 percent) reported difficulties with ADLs or IADLs or both (Altman and Bernstein, 2008). It is not surprising that policy makers and the public are confused.

Which is the real population with disabilities? Each data point can trace its origin to a conceptual component in current theoretical models, but the concepts represent different aspects of the disablement process, and the measures represent different aspects of the concepts. Researchers who have fixed their approach to disability at the level of ADLs and IADLs would probably argue that this area of focus reflects the most important health and health care issues: this is the group that needs help, this is the group that displays the most disability by displacing a greater part of the mat—from the IOM model (see Figure A-3). However, for those who work from a civil rights perspective and are attuned to the underlying intent of the Americans with Disabilities Act to make inclusion and participation available to this population, the population who are "at risk" to develop a disability because of the interaction of their functional limitation and the cultural and physical environment is the real population. It is not my place to say which measure is best, because they all serve a purpose. However, I note that the use of any one measure to the exclusion of the others distorts understanding of the total process; the coordination of the number of questions, the functions included, question wording, answer categories, and conceptual elements across surveys would help improve the confusion about numbers. And because the gaps in information are great, as demonstrated by Tables A-2 and A-3, efforts to fill those gaps would go a long way toward understanding the whole disablement process, whether the focus is a subpopulation of adults or the whole population.

How the Data Are Used

Finally, the information in Table A-5 provides a window on how the conceptual elements that are measured inform understanding of disability among the elderly. Using two journals from the past 3 years, *Demography* and the *Journal of Gerontology*, articles that used a disability measure (broadly defined) in the analyses were selected in a random manner. Table A-5 shows the article reference, the data source, the primary and secondary disability measures, and a very brief indication of how the measures were used. A total of 11 of the data sources were large national surveys, 10 in the United States and 1 from Japan. The single most frequently cited data set used was HRS, with NHIS a close second. It is interesting to note that the 1994–1995 NHIS-D accounted for most of the articles that had the NHIS as a data source.

The predominant conceptual measure used in 13 of 18 articles was a measure of ADLs and IADLs, some using a functioning perspective about difficulty and others using a behavioral perspective about receiving help. Although the conceptual base is similar, the measures vary not only by the difference between a behavioral or functional approach, but also by the number of questions involved. So, for example, the Freedman and colleagues (2007) and the Mutchler and colleagues (2007) articles use two questions, from two different data sources, to represent ADLs and IADLs (one for each concept), whereas Liang and colleagues (2008) use 11 questions to represent the ADL-IADL construct. For the remaining articles that do not use an ADL-IADL construct to represent disability, one uses a work limitation question, one uses a summed indicator of limitations in eight basic actions (also known as Nagi questions), and the final one uses the short form SF-36[7] on physical functioning and pain. This concentration of measures in a very narrow area of the disablement process reflects a very limited orientation toward disability in this population. In addition, if one were to compare the populations described in each of these articles to one another, as in Table A-4, chances are the overlap would not be extensive, nor would it be consistent.

Nine of these articles use the measure of disability as a dependent variable, and several use it to map trends in prevalence over a time period. Trends that are followed include the need for help, life expectancy, active life expectancy, and transition into and out of a disabled status (disabled status being equated with high levels of dependence). The other nine articles use the measure of disability as an independent or control variable to predict other things, such as how long baby boomers are expected to work, the

[7]The SF-36 is a comprehensive short form of a health survey, developed by John Ware, with only 36 questions that yields an eight-scale health profile as well as summary measures of health-related quality of life. For more information, see http://www.sf-36.org.

TABLE A-5 Sampling of Disability Measures and How They Are Used: *Demography* and the *Journal of Gerontology*

Article	Data Source	Primary Disability Measure
Demography		
Freedman et al. (2007)	NHIS 1997–2004 Ages 65+	Behavioral measure of ADLs/IADLs—need help [2]
Cai and Lubitz (2007)	Medicare Current Beneficiary Survey	Constructed variable combining ADL/IADL measures[a]
Mutchler et al. (2007)	2000 Census	Functioning in basic actions and self-care [3]
Journal of Gerontology		
Newcomer et al. (2005)	NHIS-D	Behavioral measure of ADLs/IADLs with indicators of unmet need in these measures
Mendes de Leon et al. (2005)	Chicago Health and Aging Project	Functioning in six ADLs, upper and lower body functioning measures
Agree et al. (2005)	NHIS-D	Count of difficulty functioning with ADL and IADL tasks
Liang et al. (2005)	Japanese Study of Older Adults	Functioning in six task items: ADLs/IADLs
Mermin et al. (2007)	HRS	Any health-related work limitation, question or questions not specified
Muramatsu et al. (2007)	HRS	Limitations with ADLs/IADLs (difficulty or help not specified)
Manton et al. (2008)	NLTCS	Behavioral measures of ADLs/IADLs (use help) and functioning in basic actions
Liang et al. (2008)	HRS	Count of difficulty functioning with ADL/IADL tasks [11]
Wolinsky et al. (2005)	African American Health Project	Five basic action measures and seven indicators of difficulty in ADLs/IADLs
Wolf et al. (2007)	New Haven EPESE	Difficulty with ADLs, functioning difficulties in basic actions[b]
Chipperfield et al. (2008)	Study of Aging in Manitoba (AIM)	Measure of functional status includes 22 items from existing ADL/IADL measures
Kasper et al. (2008)	Mothers of a study cohort composed of all first graders in Woodlawn in Chicago	Physical functioning and pain from Short Form-36—10 items
Forman-Hoffman et al. (2008)	HRS	Summed functional limitations in eight basic actions

NOTE: EPESE = Established Populations for Epidemiologic Studies of the Elderly, HRS = Health and Retirement Study, NHIS = National Health Interview Study, NHIS-D = National Health Interview Study on Disability, NLTCS = National Long-Term Care Survey.

Secondary Disability Measure	Independent or Dependent Variable
Conditions considered to cause ADL or IADL	Trend analysis—disability measure, dependent variable
None	Trend analysis—disability measure used to explain disabled life expectancy
None	Disability dependent variable predicated on immigrant status
Sensory communication Social activity	Unmet need serves as dependent variable, others as predictive variables
	Measures were used as dependent variables and examined over time by age, gender, and race
Use of assistive devices	Measures of ADLs/IADLs used as control variables, assistive device used as predictor
None	Used as independent variable to predict trajectory—referred to as functional impairment
None	Used as an independent variable to predict how long baby boomers expect to work
Conditions (used as number, not type)	Used as independent variable to predict risk of nursing home admission
One vision functioning measure and one assistive device measure [Total = 27]	Measures used to construct seven dimensions of disability to predict active life expectancy
Conditions, depressive symptoms	Used as dependent variable—predicting changes associated with age and gender
None	Dependent variable predicted in part by previous subclinical status
None	Trend of transition into and out of disabled status
Also used chronic condition list	Used as independent variable to predict physical activity
Health conditions—list Condition limits activity in any way	Dependent variable associated with poverty and family stress
Depressive symptoms and medical comorbidity based on conditions	Used as independent variable to predict weight changes

[a]It is not indicated in the paper if the ADL/IADL measure is taken from the functioning questions about difficulties or the behavioral question about needing help.

[b]Measures are not organized on any theoretical structure but are based on Katz, Rostow, and Nagi measures.

risk of nursing home admission, the use of assistive devices, physical activity, and weight changes. In a few instances, there are indications of either a lack of familiarity with the current theoretical models of disability or lack of understanding of the value of some of the measures that are available to be used. The first problem can be found in several of the articles in the use of terminology to discuss disability aspects of the analysis. A common terminology still is not being used and in some cases the conceptual elements of theoretical disability models (identifying functioning difficulty as impairment) are being confused.

Another problem is the continued restrictive interpretation of disability measures available for use and the lack of appreciation of broader aspects of disability (as conceptualized beyond the dependence focus that has been demonstrated here and inclusive of environmental factors when available). An example is the Mermin and colleagues (2007) article, which examines predictors of the length of time baby boomers feel they will work before retirement. HRS is used and an indicator of work limitation is one of the independent variables. It is a routine assumption that persons with work limitations will not work as long as persons without work limitations. However, there are available data in HRS that could have been used to test this routine assumption. HRS has a series of 10 questions that indicate if and how an employer has provided help so that a worker can stay on the job. If that information had been factored into the analysis, it could have given some insight into the effect of environmental factors on disability behavior associated with retirement and provided a very meaningful comparison related to retirement choices.

As with the examination of the data elements of the commonly used surveys in Tables A-2 and A-3, the literature reviewed hints at an even narrower use of the data for research purposes. Available environmental elements are used in only one of the articles, and that article focuses on tracking the use of assistive devices. The various role participation representations were used in only 1 of the 18 analyses. The broad emphasis of the literature seems to be descriptive of the status of dependence, either documenting changes in the prevalence of the disability variable over time or the effect of various other statuses—immigrant, age, race, gender—on prevalence of dependence (as represented by the ADL-IADL measures). Of course, using such a small sample of articles from only two journals provides very little data on which to make a fully accurate assessment.

CONCLUSIONS

Although there are some very useful theoretical models of disability to underpin a wide-ranging research agenda about disability, the measurement of the conceptual components in the existing data sets and the use of data already available have not yet reached their potential. In some instances, the data and the analyses are still too closely tied to the medical conceptualization of disability, as a problem located within the individual. In other instances, the reduced financing and increased demand on data collection processes have forced choices that limit the expansion of the conceptual coverage of the data. Integral components of the disability process, elements of role participation, and the interaction of the individual and the environment are vital to a more complete understanding of disability, but they are not generally considered.

Although groups at various universities are working on some of these measurement problems, there is no general funded mandate, particularly in government-sponsored data collection efforts, to test or expand disability measurement beyond what is currently in use. Nor, until this year, has there been any coordination across data collection entities to consider some form of standardization of core efforts. A recent hearing in the U.S. House of Representatives (in June 2009), held in response to the National Council on Disability (2008) publication about disability data, may have drawn some more serious attention to this issue. The testing and approval of the ACS disability questions and their adoption for use in CPS, in the National Crime Victimization Survey for the Department of Justice, and for testing in NHIS may foretell the eventual closer coordination of disability measures, if not their immediate expansion.

Jette and Badley (2002, p. 183) have noted that the "field of disability research is in need of uniform concepts and a common language to guide scholarly discussion, to advance theoretical work on the disablement process, to facilitate future survey and epidemiological research and to enhance understanding of disability on the part of professionals as well as the general public." I strongly agree and would add that the field also needs some expansion of measurement to cover all the conceptual components of our theories, as well as coordination of measures that represent those theoretical concepts. A very careful consideration of the use and interpretation of measures, particularly in research that is used to inform policy, is critically necessary. Whatever uniformity is developed in conceptualization and language is still undermined when it is not used in creating operationalized definitions of concepts and there are multiple variations in question-and-answer language.

The strong medical model basis of work on disability in gerontology has created a rich set of measures that are narrowly focused in the area of identifying behavioral and functioning dependence. The measures have provided important policy information about problems and cost of the final one-eighth to one-tenth of the life-cycle. The data based on those measures have documented changes in longevity and health in people's final years and have provided information that has allowed for the development of assistance programs that go beyond warehousing our elders in institutions. However, the very data and forms of measurement that have helped identify and understand the change in the structure of the problems of the aging population are no longer sufficient to address all the issues, particularly in relation to role participation. A 65-year-old person who is facing the very likely possibility of another 20 years of life may also experience the role changes associated with gradual development of functional limitations in basic actions before reaching the stage of dependence. Understanding the issues associated with the organized activities that are key to maintaining role participation, including the role that society and environment play in the process, can go a long way to improving those 20 years, as well as the final period of dependence.

Finally, it is interesting that while none of the models currently in use proposes differentiating the meaning of disability among the three major age groups—children, adults, and the elderly—we go about measuring disability within those groups in very different ways. Children's disability is measured using either conditions or use of special services; adults' disability is measured most broadly through a combination of functioning in basic actions, limitations in work, and ADLs and IADLs, with an emphasis on work limitation; and the aging population's disability is represented most commonly by behavioral or functioning representations of ADLs and IADLs or, somewhat less frequently, by levels of functioning in basic actions.

Ferraro (2006, p. S3) has made the point that "gerontology flirts with being medicalized" and should investigate topics not directly related to health. Studying the problems of disability from a role participation perspective would move away from that health focus and contribute to the overall understanding of the aging process over the life-cycle. Expanding survey measurement to include a fuller, more robust coverage of functioning in basic actions, indicators of organized activity participation meaningful to role participation, and measures of key environmental factors would facilitate the study of disability across all age groups.

ANNEX

AMERICAN COMMUNITY SURVEY QUESTIONS, DEFINITION OF TERMS, AND SUGGESTED ADDITIONS TO QUESTIONS

American Community Survey Questions[8]

Answer question 16 ONLY IF this person is 1 year old or older. Otherwise skip to the questions for person 2 on page 12.

16. a. Is this person deaf or does he/she have serious difficulty hearing? YES/NO
    b. Is this person blind or does he/she have serious difficulty seeing even when wearing glasses? YES/NO

Answer question 17a ONLY IF this person is 5 years old or older. Otherwise, skip to the questions for person 2 on page 12.

17. a. Because of a physical, mental, or emotional condition, does this person have serious difficulty concentrating, remembering, or making decisions? YES/NO
    b. Does this person have serious difficulty walking or climbing stairs? YES/NO
    c. Does this person have difficulty dressing or bathing? YES/NO

Answer 18 ONLY IF this person is 15 years old or older. Otherwise skip to the questions for person 2 on page 12.

18. Because of a physical, mental, or emotional condition, does this person have difficulty doing errands alone such as visiting a doctor's office or shopping?

Definition of Terms

Willful action—Based on either performance or capacity, action reflects the individual's will to carry out basic volitional bodily operations at the level of the organism (whole person). Examples include walking, climbing steps, reading, communicating, etc. It is distinct from body functions (ICF), which are "physiological functions of body systems" rather than functions of the whole person. When combined, multiple actions can result in performance of tasks (Nordenfelt, 2003). In the ICF, actions are included in the domain

[8]This section is excerpted from the full ACS questionnaire, which is available at http://www.census.gov/acs/www/Downloads/SQuest09.pdf.

of activity defined as "the execution of a *task or action* by an individual, representing the individual perspective of functioning" (World Health Organization, 2001b, p. 190). ICF does not differentiate actions and tasks.

Specific task—Execution of a group of willful actions by an individual. It is an indicator of a series of related or more complicated actions necessary to accomplish an objective, which is a central component of role behavior. Examples include bathing, dressing, and feeding, which are central elements of self-care, or driving a car and planning a meeting, which can be central elements of employment. In ICF, tasks are included in the domain of activity defined as "the execution of a *task or action* by an individual, representing the individual perspective of functioning" (World Health Organization, 1991b, p. 190). ICF does not differentiate actions and tasks.

Organized activity—Represents the accomplishment of a variety of specific tasks and willful actions in order to complete an activity that is socially recognized or defined in a culture. An example would be going out to dinner, which entails making reservations, getting dressed appropriately, finding transportation, engaging with friends, reading a menu, ordering, paying the bill, leaving a tip, and other details.

Role participation—Represents the accomplishment, through willful actions, specific tasks and organized activities, of enough elements of a social role to claim that form of role participation as represented in a particular culture or society.

Current Questions and Suggested Additions

National Health and Nutrition Examination Survey

In a typical week, how many times do you talk on the telephone with family, friends or neighbors?

How often do you get together with friends or relatives; I mean things like going out together or visiting in each other's homes? (per year)

About how often do you visit with any of your neighbors, either in their homes or in your own?

How often do you attend church or religious services?

Do you belong to any clubs or organizations such as church groups, unions, fraternal or athletic groups, or school groups?

Altogether how often do you attend meetings of the clubs or organizations to which you belong?

Suggested additions to monitor behavioral changes would include some measurement of how this behavior differs from that which occurred before the onset of the limitation. It would require tracking of what has been reported as limitations in functioning or tasks or organized activities.

An example would be additions to the following questions to track those changes:

Do you belong to any clubs or organizations such as church groups, unions, fraternal or athletic groups, or school groups?

Have your group memberships increased, decreased, or remained the same since the onset of [fill in with type of functional limitation, i.e., your experience of limitation in walking—various ways to include limitations in different areas of physical functioning can be developed]?

Altogether how often do you attend meetings of the clubs or organizations to which you belong?

Has your meeting attendance increased, decreased, or remained the same since the onset of [fill in with type of functional limitation, i.e., your experience of limitation in walking]?

Next, depending on the answer provided (increase, decrease, or remained the same) a follow-up could probe why the behavior changed (increased or decreased) with no questions if it stayed the same.

Longitudinal Study of Aging

During the past two weeks, did you—

Get together socially with friends or neighbors?
Talk with friends or neighbors on the telephone?
Get together with ANY relatives not including those living with you?
Go to church, temple or another place of worship for the services or other activities?
Go to a show or movie, sports event, club meeting, class or other group event?
Go out to eat at a restaurant?
Do unpaid volunteer work, such as teaching, coaching, office work, or providing care?
Participate in Elderhostel?

Suggested additions would be similar to those suggested above. The operative element being to discover if this activity represents a change from the activity level or type of activity experienced before the functional limitations began to be noticeable.

REFERENCES

Adler, M. (1989). *Task 1: Population Profile of Disability.* Washington, DC: U.S. Department of Health and Human Services.

Agree, E.M., Freedman, V.A., Cornman, J.C., Wolf, D.A., and Marcotte, J.E. (2005). Reconsidering substitution in long-term care: When does assistive technology take the place of personal care? *Journal of Gerontology: Social Sciences* 60B(5):S272–S280.

Altman, B.M. (1984). *Primary, Secondary and Tertiary Factors Influencing Disability.* Unpublished dissertation. Sociology Department, University of Maryland, College Park.

Altman, B.M. (2001a). Definitions of disability and the operationalization and measurement in survey data: An update. In S.N. Barnartt and B.M. Altman (Eds.), *Exploring Theories and Expanding Methodologies: Where We Are and Where We Need to Go* (pp. 77–100). London, England: Elsevier.

Altman, B.M. (2001b). Disability definitions, models, classification schemes, and applications. In G. Albrecht, K. Seelman, and M. Bury (Eds.), *Handbook of Disability Studies* (pp. 97–22). Thousand Oaks, CA: Sage.

Altman, B.M., and Bernstein, A. (2008). *Disability and Health in the United States, 2001–2005.* Hyattsville, MD: National Center for Health Statistics.

Altman, B.M., Rasch, E.K., and Madans, J.H. (2006). Disability measurement matrix: A tool for the coordination of measurement purpose and instrument development. In B.M. Altman and S.N. Barnartt (Eds.), *International Views on Disability Measures: Moving Toward Comparative Measurement* (pp. 263–284). Amsterdam, The Netherlands: Elsevier.

Bogdan, R., and Biklen, D. (1977). Handicapism. *Social Policy* 15:537–550.

Brault, M., Stern, S., and Raglin, D. (2007). *Evaluation Report Covering Disability.* (2006 American Community Survey Content Test Report P.4.) Washington, DC: U.S. Census Bureau. Available: http://www.census.gov/acs/www/AdvMeth/content_test/P4_Disability.pdf [accessed July 2008].

Cai, L., and Lubitz, J. (2007). Was there compression of disability for older Americans from 1992 to 2003? *Demography* 44(3):479–495.

Caplan, A. (1988). Is medical care the right prescription? In S. Sullivan and M.E. Lewin (Eds.), *The Economics and Ethics of Long-Term Care and Disability*. Washington, DC: American Enterprise Institute for Public Policy Research.

Chapireau, F. (2005). The environment in the international classification of functioning, disability and health. *Journal of Applied Research in Intellectual Disabilities* 16(4):305–311.

Chipperfield, J.G., Newall, N.E., Chuchmach, L.P., Swift, A.U., and Haynes, T.L. (2008). Differential determinants of men's and women's everyday physical activity in later life. *Journal of Gerontology: Social Sciences* 63B(4):S211–S218.

Ferraro, K.F. (2006). Is gerontology inclusive? *Journal of Gerontology: Social Sciences* 61B(1): S3.

Ferron, D.T. (1981). *Disability Survey 72: Disabled and Nondisabled Adults, A Monograph.* (Research Report No. 56, SSA Publication No. 13-11812.) Washington, DC: U.S. Department of Health and Human Services.

Forman-Hoffman, V.L., Richardson, K., Yankey, J.W., Hillis, S.L., Wallace, R.B., and Wolinsky, F.F. (2008). Retirement and weight changes among men and women in the Health and Retirement Study. *Journal of Gerontology: Social Sciences* 63B(3):S146–S153.

Freedman, V.A., Crimmins, E., Schoeni, R.F., Spillman, B.C., Aykan, H., Kramarow, E., Land, K.C., Lubitz, J., Manton, K.G., Martin, L.G., Shinberg, D., and Waidmann, T. (2004). Resolving inconsistencies in trends in old-age disability: Report from a technical working group. *Demography* 42(3):417–441.

Freedman, V.A., Schoeni, R.F., Martin, L.G., and Cornman, J.C. (2007). Chronic conditions and the decline in late-life disability. *Demography* 44(3):459–477.

Harris, H.B., Hendershot, G., and Stapleton, D.C. (2005). *A Guide to Disability Statistics from the National Health Interview Survey.* Rehabilitation Research and Training Center on Disability Demographics and Statistics, Cornell University. Available: http://digitalcommons.ilr.cornell.edu.cgi/viewcontent.cgi?filename=0&article=1184&context=edicollect&type=additional [accessed June 2009].

Institute of Medicine. (1991). *Disability in America: Toward a National Agenda for Prevention.* Committee on a National Agenda for the Prevention of Disabilities. A.M. Pope and A.R. Tarlov (Eds.). Washington, DC: National Academy Press.

Institute of Medicine. (1997). *Enabling America: Assessing the Role of Rehabilitation Science and Engineering.* Committee on Assessing Rehabilitation Science and Engineering. E.N. Brandt and A.M. Pope (Eds.). Washington, DC: National Academy Press.

Institute of Medicine. (2007). *The Future of Disability in America.* Committee on Disability in America. M.J. Field and A.M. Jette (Eds.). Washington, DC: The National Academies Press.

Jette, A.M., and Badley, E. (2002). Conceptual issues in the measurement of work disability. In Institute of Medicine and National Research Council, *The Dynamics of Disability: Measuring and Monitoring Disability for Social Security Programs.* G.S. Wunderlich, D.P. Rice, and N.L. Amado (Eds.). Committee to Review the Social Security Administration's Disability Decision Process Research. Washington, DC: National Academy Press.

Kasper, J.D., Ensminger, M.E., Green, K.M., Fothergill, K.E., Juon, H., Robertson, J., and Thorpe, R.J. (2008). Effects of poverty and family stress over three decades on the functional status of older African American women. *Journal of Gerontology: Social Sciences* 63B(4):S201–S210.

Katz, S., Ford, A.B., Moskowitz, R.W., Jackson, B.A., and Jaffee, M.W. (1963). Studies of illness in the aged: The index of ADL, a standardized measure of biological and psychosocial function. *Journal of the American Medical Association* 185:914–919.

Liang, J., Shaw, B.A., Krause, N., Bennett, J.M., Kobayashi, E., Fukaya, T., and Sugihara, Y. (2005). How does self-assessed health change with age? A study of older adults in Japan. *Journal of Gerontology: Social Sciences* 60B(4):S224–S232.

Liang, J., Bennett, J.M., Shaw, B.A., Quinones, A.R., Ye, W., Xu, X., and Ofstedal, M.B. (2008). Gender differences in functional status in middle and older age: Are there any age variations? *Journal of Gerontology: Social Sciences* 63B(5):S282–S292.

Madans, J., Altman, B., Rasch, E., Mbogoni, M., Synneborne, M., Banda, J., Me, A., and DePalma, E. (2004). *Washington City Group Position Paper: Purpose of Measurement.* Hyattsville, MD: National Center for Health Statistics. Available: http://www.cdc.gov/nchs/about/otheract/citygroup/meeting3.htm#papers [accessed October 2008].

Manton, K.G., Gu, X., and Lowrimore, G.R. (2008). Cohort changes in active life expectancy in the U.S. elderly population: Experience from the 1982–2004 National Long-Term Care Survey. *Journal of Gerontology: Social Sciences* 63B(5):S269–S281.

Mendes de Leon, C.F., Barnes, L.L., Bienias, J.L., Skarupski, K.A., and Evans, D.A. (2005). Racial disparities in disability: Recent evidence from self-reported and performance-based disability measures in a population-based study of older adults. *Journal of Gerontology: Social Sciences* 60B(5):S263–S271.

Mermin, G.B., Johnson, R.W., and Murphy, D.P. (2007). Why do boomers plan to work longer? *Journal of Gerontology: Social Sciences.* 62B(5):S286–S294.

Muramatsu, N., Yin, H., Campbell, R.T., Hoyem, R.L., Jacob, M.A., and Ross, C.O. (2007). Risk of nursing home admission among older Americans: Does states' spending on home and community-based services matter? *Journal of Gerontology: Social Sciences* 62B(3): S169–S178.

Mutchler, J.E., Prakash, A., and Burr, J.A. (2007). The demography of disability and the effects of immigrant history: Older Asians in the United States. *Demography* 44(2):251–263.

Nagi, S. (1965). Some conceptual issues in disability and rehabilitation. In M. Sussman (Ed.), *Sociology and Rehabilitation* (pp. 100–113). Washington, DC: American Sociological Association.

Nagi, S. (1969). *Disability and Rehabilitation: Legal, Clinical, and Self-Concepts and Measurement.* Columbus: Ohio University Press.

Nagi, S. (1991). Disability concepts revisited: Implications for prevention. In Institute of Medicine, *Disability in America: Toward a National Agenda for Prevention* (pp. 309–327). Committee on a National Agenda for the Prevention of Disabilities. A.M. Pope and A.R. Tarlov (Eds.). Washington, DC: National Academy Press.

National Center for Health Statistics. (1959). *Impairments by Type, Sex, and Age.* (National Health Interview Survey.) Washington, DC: U.S. Department of Health, Education, and Welfare.

National Council on Disability. (2008). *Keeping Track: National Disability Status and Program Performance Indicators.* Washington, DC: Author. Available: http://www.ncd.gov/newsroom/publications/2008/pdf/Indicators_Report.pdf [accessed May 2008].

Newcomer, R., Kang, T., LaPlante, M., and Kaye, S. (2005). Living quarters and unmet personal care assistance among adults with disabilities. *Journal of Gerontology: Social Sciences* 60B(4): S205–S213.

Nordenfelt, L. (1995). On the general concepts of action and ability. *International Journal of Technology Assessment in Health Care* 11(2):144–152.

Nordenfelt, L. (2003). Disability, action theory and ICF. *Newsletter on the World Health Organization Family of International Classifications* 1(1):13–15.

Perrott, G., Tibbitts, C., and Britten, R. (1939). *The National Health Survey: Scope and Method of the Nation-Wide Canvas of Sickness in Relation to Its Social and Economics Setting.* Washington, DC: U.S. Public Health Service.

Rasch, E., Hochberg, M., Magder, L., Magaziner, J., and Altman, B.M. (2008). Health of community-dwelling adults with mobility limitations in the United States: Incidence of secondary health conditions. Part II. *Archives of Physical Medicine and Rehabilitation* 89:219–230.

Silverstein, R., Julnes, G., and Nolan, R. (2005). What policy makers need and must demand from research regarding the employment rate of persons with disabilities. *Behavioral Sciences and the Law* 23:399–448.

Thomas, C. (2002). Disability theory: Key ideas, issues and thinkers. In C. Barnes, M. Oliver, and L. Barton (Eds.), *Disability Studies Today* (pp. 38–57). Malden, MA: Blackwell.

Verbrugge, L.M., and Jette, A.M. (1994). The disablement process. *Social Science and Medicine* 38:1–14.

Whiteneck, G. (2006). Conceptual models of disability: Past, present, and future. In Institute of Medicine, *Disability in America: A New Look. Summary and Background Papers* (pp. 50–66). M.J. Field, A.M. Jette, and L. Martin (Eds.). Washington, DC: The National Academies Press.

Wolf, D.A., Mendes de Leon, C.F., and Glass, T.A. (2007). Trends in rates of onset of and recovery from disability in older ages: 1982–1994. *Journal of Gerontology: Social Sciences* 62B(1):S3–S10.

Wolinsky, F.D., Miller, D.K., Andresen, E.M., Malmstrom, T.K., and Miller, J.P. (2005). Further evidence for the importance of subclinical functional limitation and subclinical disability assessment in gerontology and geriatrics. *Journal of Gerontology, Social Sciences* 60B(3):S146–S151.

World Health Organization. (1980). *International Classification of Impairments, Disabilities, and Handicaps.* Geneva, Switzerland: World Health Organization.

World Health Organization. (2001a). *International Classification of Impairments, Disabilities, and Handicaps.* Geneva, Switzerland: World Health Organization.

Appendix B

Workshop Agenda and Presenters

Workshop on Improving Measurement of Late-Life Disability in Population Surveys: Beyond ADLs and IADLs

January 8–9, 2009

National Academies Keck Center, Room 100
500 Fifth Street NW, Washington, DC

Agenda

January 8, 2009

OPENING SESSION

9:00–10:15 a.m.	Welcome and Introductions	Alan Jette, *Chair*
	Welcoming Remarks on Behalf of the National Academies	Michael Feuer
	Welcome and Sponsor's Perspectives	Richard Suzman
9:45–10:15 a.m.	Background and Context of the Workshop	Alan Jette
10:15 a.m.	*Break*	
10:30 a.m.–12:30 p.m.	**SESSION ONE**	***Moderator:*** Vicki Freedman
	Challenges to Improve Current Measurement of Late-Life Functioning and Disability in Population Surveys (*taking into account suitability for population surveys, relevance for monitoring trends, response burden and bias, and costs)*:	

		Presenters:
10:30–10:45 a.m.	Overview of Background Paper— Population Survey Measures of Functioning: Strengths and Weaknesses	Barbara Altman
10:45–11:00 a.m.	Developing Questions in Surveys to Identify People Early in the Disablement Process	Linda Fried
11:00–11:15 a.m.	Enhancing the Ascertainment of Disability	Thomas Gill
11:15–11:30 a.m.	Self Versus Proxy Responses in Population Surveys	Jay Magaziner
11:30–11:45 a.m.	Expanding Mode of Survey Administration – (face to face, telephone, Internet, etc.)	Arie Kapteyn
11:45 a.m.– 12:15 p.m.	Comments and General Discussion	
12:15 p.m.	*Working Lunch—Continuation of Discussion*	
1:30–3:00 p.m.	**SESSION TWO**	***Moderator:*** Alan Jette
	Potential Methods for Refining or Augmenting Current Measures of Late-Life Disability in Population Surveys to Foster Comparability Across Key Subgroups (*taking into account suitability for population surveys, relevance for monitoring trends, response burden and bias, and costs*)	
		Presenters:
1:30–1:45 p.m.	Performance Measures in Population Surveys: • Initial use and diffusion • Using performance measures to calibrate disability cut-points across population groups	Jack Guralnik
1:45–2:00 p.m.	Improving Patient Reported Measurement of Disability Using Item Response Theory/Computer-Adaptive Testing Techniques: • Are these suitable for population surveys? • Are they relevant for monitoring trends?	Karon Cook

	• Capture changes in functionality across thresholds.	
2:15–2:30 p.m.	Developing Measures of Time Use to Study Disability in the Panel Study of Income Dynamics; Potential for Using Time-Use Data to Augment Existing Measures of ADLs and IADLs: • What elements of time use should be measured? • Are these suitable for population surveys and relevant for monitoring trends?	Vicki Freedman
2:30–3:00 p.m.	Comments and General Discussion	
3:00 p.m.	*Break*	
3:30–5:00 p.m.	**SESSION THREE**	***Moderator:*** Linda Martin
	Improving the Validity of Cross-Population Comparisons—within the United States and Across Countries—for Measures of Disability Obtained in Population Surveys: • Can vignettes validate judgmental reports in population surveys? • Identification of homogeneous groups (demographic/medical) to improve comparisons • Cross-cultural comparisons	
		Presenters:
3:3–3:45 p.m.	Cognitive Limitation of Functioning and Disability—Additional Measures That Could Be Used in Surveys and Their Relationship to Current Measures	Craig Velozo
3:45–4:00 p.m.	Using Vignettes to Improve Cross-Population Comparability of Self-Rated Disability Measures in Population Surveys	Arthur van Soest
4:00–4:15 p.m.	New Approaches to Cognitive and Field Testing of Disability Measures for Improving Cross-Cultural Comparability	Julie Weeks
4:15–5:00 p.m.	Comments and General Discussion	
5:00–6:00 p.m.	*Reception for All Attendees*	

January 9, 2009

9:00 a.m.	**OPENING REMARKS** *(Continental breakfast available in meeting room)*	Alan Jette
9:00–10:15 a.m.	**SESSION FOUR**	***Moderator:*** Linda Fried
	The Need to Measure Functioning and Disability in Context (*taking into account suitability for population surveys, relevance for monitoring trends, response burden and bias, and costs, including* • environmental barriers and adaptations, • changes/differences in standards in what is acceptable, and • pychological adaptations.	
		Presenters:
9:00–9:15 a.m.	Incorporating Assistive Technology and Home Modification Measures in Population Surveys	Emily Agree
9:15–9:30 a.m.	Incorporating Questions on Behavioral Adaptation to Functional Limitation Measures in Population Surveys	Carlos Weiss
9:30–9:45 a.m.	The Utility of Participation Measures in Population Surveys	Gale Whiteneck
9:45–10:15 a.m.	Comments and General Discussion	
10:15 a.m.	*Break*	
10:30 a.m. – noon	**SESSION FIVE**	***Moderator:*** Alan Jette
	Priority Areas for Research and Development Toward Improved Measures of Late-Life Disability in Population Surveys	
	➢ Panel of three key participants will lead off the discussion ➢ Discussion and comments from participants and attendees ➢ General discussion on disability measures	Thomas Gill Arthur van Soest David Weir

WORKSHOP PRESENTERS

Emily Agree is an associate professor and associate director of the Hopkins Population Center at the Bloomberg School of Public Health, Johns Hopkins University.

Barbara Altman is a disability statistics consultant.

Karon Cook is a research associate professor at the University of Washington in Seattle.

Michael Feuer is executive director of the Division of Behavioral and Social Sciences and Education, National Research Council.

Vicki A. Freedman is a professor in the Department of Health Systems and Policy at the School of Public Health, University of Medicine and Dentistry of New Jersey.

Linda Fried is dean of the Mailman School of Public Health and professor of epidemiology and medicine and senior vice president of Columbia University Medical Center.

Thomas Gill is a professor of medicine, epidemiology, and investigative medicine at the School of Medicine and at the Adler Geriatric Assessment Center of Yale University.

Jack Guralnik is chief of the Laboratory of Epidemiology, Demography, and Biometry at the National Institute on Aging, National Institutes of Health.

Alan M. Jette is director, Health & Disability Research Institute, and professor, Health Policy and Management, at Boston University, School of Public Health.

Arie Kapteyn is program director in labor and population at the RAND Corporation.

Jay Magaziner is professor and chair of the Department of Epidemiology and Preventive Medicine, School of Medicine, at the University of Maryland, Baltimore.

Linda G. Martin is senior fellow at the RAND Corporation.

Richard Suzman is associate director for Behavioral and Social Research at the National Institute on Aging, National Institutes of Health.

Arthur van Soest is an economist in the Department of Econometrics & OR at Tilberg University, Tilburg, Netherlands, and at the RAND Corporation.

Craig Velozo is in the Department of Occupational Therapy at the College of Public Health and Health Professions at the University of Florida and the Veterans Affairs Medical Center in Gainsville, Florida.

Julie D. Weeks is acting chief of the Aging and Chronic Disease Studies Branch and project director of longitudinal studies of aging in the Office of Analysis and Epidemiology of the National Center for Health Statistics, Centers for Disease Control and Prevention.

David Weir is a research professor at the Survey Research Center at the University of Michigan.

Carlos Weiss is assistant professor in the Division of Geriatric Medicine at the Johns Hopkins University School of Medicine.

Gale Whiteneck is director of research at Craig Hospital in Englewood, Colorado.

Appendix C

Biographical Sketches of Steering Committee Members

Alan M. Jette (*Chair*) directs the Health and Disability Research Institute and is a professor of health policy and management at the School of Public Health, both at Boston University. He has been active in reviews of the Social Security Administration's disability decision process research. His research emphases include late-life exercise; evaluation of treatment outcomes; and the measurement, epidemiology, and prevention of late-life disability. He has published more than 125 articles on these topics in the rehabilitation, geriatrics, and public health literature. He has an M.P.H. degree in health gerontology and a Ph.D. degree in public health behavior, both from the University of Michigan.

Vicki A. Freedman is a professor of health systems and policy at the University of Medicine and Dentistry of New Jersey's School of Public Health. She is a demographer and chronic disease epidemiologist who specializes in the measurement of disabilities in older populations. She has published extensively on the topics of population aging, disability, and long-term care, including several widely publicized articles on trends in late-life functioning. Her current research emphasizes interventions that can be used to prevent late-life disability decline, the socioeconomic and racial disparities in the incidence of late-life disabilities, the causes of late-life disability trends, and the role of assistive technology in ameliorating disability. She has served on more than a dozen national advisory panels for federal agencies, including the National Institute on Aging and the U.S. Department of Health and Human Services. She earned her Ph.D. in epidemiology from Yale University and M.A. in demography from Georgetown University.

Linda P. Fried is dean of the Mailman School of Public Health and senior vice president of Columbia University Medical Center. Her core research interests are prevention and health promotion for older adults, with particular emphasis on the discovery of the causes of frailty and disability and their prevention. She is a board-certified internist and geriatrician, with postdoctoral training at the Johns Hopkins Medical Institutions in geriatrics, general internal medicine, and epidemiology (cardiovascular and aging). She is a recipient of a National Institute on Aging MERIT Award. She is a member of the Institute of Medicine. She received an M.D. degree from Rush Medical College and an M.P.H. degree from the Johns Hopkins University School of Hygiene and Public Health.

Linda G. Martin is a senior fellow at RAND, serving as an advisor on a variety of RAND studies and activities, as well as conducting her own research on the health of older people in the United States and Asia. She is also an adjunct professor in the Department of Population, Family, and Reproductive Health at the Bloomberg School of Public Health at Johns Hopkins University. Previously she served as president of the Population Council; as vice president for research development at RAND; and as director of the Committee on Population at the National Research Council. She has held other teaching and research appointments at Princeton University, Georgetown University, and the University of Michigan. She has M.P.A. and Ph.D. degrees from Princeton University.

Joshua Salomon is an associate professor of international health in the Department of Population and International Health at Harvard University. His research focuses on priority setting in global health in three main substantive areas: measurement of population health status and health valuations; modeling and forecasting health outcomes and disease burden; and evaluation of the potential impact and cost-effectiveness of current and future health interventions. He is an investigator on projects funded by National Institute on Aging and the Gates Foundation relating to summary measures of population health; modeling HIV/AIDS epidemics and interventions for prevention and treatment; modeling disease outcomes for population health monitoring and surveillance; and evaluating the potential impact and cost-effectiveness of new vaccines. He also leads a collaborative project with the Mexican Ministry of Health on priority setting for interventions in the context of health reform. He has a Ph.D. degree from Harvard University.

Arthur A. Stone is a distinguished professor of psychiatry and of psychology, vice chair of the Department of Psychiatry and Behavioral Sciences, and director of the Applied Behavioral Medicine Research Institute at Stony Brook University. His research interests focus on self-reports of medical and psychological outcomes, ecological momentary assessment, stress and coping, psychoendocrinology, and behavioral medicine. He has received several awards for his work and held editorial positions on several peer-reviewed journals. He has a Ph.D. degree in psychology (clinical) from the State University of New York at Stony Brook.

COMMITTEE ON POPULATION

The Committee on Population was established by the National Research Council in 1983 to bring the knowledge and methods of the population sciences to bear on major issues of science and public policy. The committee's work includes basic studies of fertility, health and mortality, and migration aimed at improving programs for the public health and welfare in the United States and developing countries. The committee also fosters communication among researchers in different disciplines and countries and policy makers in government, international agencies, and private organizations. The work of the committee is made possible by funding from several government agencies and private foundations.

COMMITTEE ON NATIONAL STATISTICS

The Committee on National Statistics (CNSTAT) was established in 1972 at the National Academies to improve the statistical methods and information on which public policy decisions are based. The committee carries out studies, workshops, and other activities to foster better measures and fuller understanding of the economy, the environment, public health, crime, education, immigration, poverty, welfare, and other public policy issues. It also evaluates ongoing statistical programs and tracks the statistical policy and coordinating activities of the federal government, serving a unique role at the intersection of statistics and public policy. The committee's work is supported by a consortium of federal agencies through a National Science Foundation grant.